# General Radiotelephone Operator's License Study Guide

## Third Edition

To my wife, Karen,
for encouraging me to write this book,
and
to Janette, Tammy, and Steve.

# General Radiotelephone Operator's License Study Guide

### Third Edition

## Thomas LeBlanc
## NX7P

**TAB BOOKS**

Blue Ridge Summit, PA

**Notices**

Teflon™    E.I. Dupont de Nemours, Wilmington, DE.


THIRD EDITION
FIRST PRINTING

**Library of Congress Cataloging-in-Publication Data**

LeBlanc, Thomas.
    General radiotelephone operator's license study guide / by Thomas
LeBlanc.—3rd ed.
        p.   cm.
    Includes index.
    ISBN 0-8306-3555-6 (h)   ISBN 0-8306-3554-8 (p)
    1. Radiotelephone—United States—Examinations—Study guides.
    2. Radio operators—Licenses—United States.   I. Title.
    TK6554.5.L42   1992
    621.3845′076—dc20                                                  91-47506
                                                                            CIP

TAB Books offers software for sale. For information and a catalog, please
contact TAB Software Department, Blue Ridge Summit, PA 17294-0850.

Acquisitions Editor: Roland S. Phelps
Managing Editor: Sandra Johnson Bottomley
Book Editor: B.J. Peterson
Director of Production: Katherine G. Brown
Book Design: Jaclyn J. Boone
Paperbound Cover: Holberg Design, York, PA                              EL1

# Contents

# Foreword
# to the first edition

When Tom LeBlanc asked me to write a foreword for his book, I was quite flattered. I am a Senior Electronics Instructor for ITT Technical Institute in Boise, Idaho, and have taught a number of FCC License preparation courses. In the process of teaching these courses, I had assembled a great deal of material and had an excellent success rate—until the FCC test changed in November, 1986.

Most people agreed that a new, more relevant test was needed, but the new General Radiotelephone Operator License exam contains so much new material that the old study guides, and certainly the material I used in my FCC classes, were of little value. I had to start developing new study material, which was difficult because my experience was with the old version of the test. It was at this point that I learned about Tom's book, and we became friends. Knowing the study guide addressed the new version of the test that Tom took and passed, I adopted it as a text for my FCC license preparation class. I think this book is exactly what was needed. It provides a general review, as well as the new, specialized information that is lacking in all other existing study guides.

I believe in certification because it pits the individual against himself and his knowledge against that of the experts. The FCC General Radiotelephone Operator License exam is not easy. It requires hard work, study, and guidance. I think Tom's book will provide the information and guidance that is needed to pass the test. I am a Master Certified Electronics Technician with the Electronic Technicians Association International, a NARTE Certified Engineer, a CET with ISCET, and, of course, I have a General Radiotelephone Operator License with Radar Endorsement (old test), so I know a little something about industry certification tests.

Certification tests are designed to be valid gauges by which an individual's abilities can be judged and measured. I think that anyone who takes the FCC exam should also consider becoming certified by, and active in, one of the professional organizations of the electronics industry. I think Tom's book will be of value in reviewing for most examinations in the communications field, even though it was designed with the FCC test in mind. I have already noticed that my students who have studied Tom's book score higher on the CET test.

I want to wish Tom the best for his new book, and you, the reader, success with your exam. To the reader: If you do not make it the first time, study some more, and try again; if it were easy, everyone would have an FCC license and it would not mean a thing.

Glenn M. Brusse
ITT Technical Institute
Boise, Idaho

# Foreword
# to the second edition

A great deal has transpired since Tom's first edition was released. The book has been expanded three times. The first edition, released in July of 1987, was good. This second edition promises to be the best. This book works. If you want or need an FCC General Radiotelephone License, you are reading the right book. My own experience with the book has been nothing short of phenomenal. My last class, January 1988, had a 75% pass rate. National pass rate is under 5%. In August of 1987, 68 people took the General Radiotelephone test in Portland, Oregon. Eleven passed. All eleven used Tom's book. If you want this license, use this book. It has all the information you need to know.

Glenn M. Brusse
ITT Technical Institute
Boise, Idaho

# Preface
# to the third edition

In November of 1986, the FCC released a series of "revised" exams for the General Radiotelephone Operator License. Tube theory that permeated the old series of exams was replaced with transistor and digital theory. Old rules and regulations were replaced with new marine and aviation rules and regulations. Perhaps 30 – 40% of old material remained in the test question pool. The FCC's examination revision was so extensive that all existing Q&A's became instantly obsolete. As a result of inadequate study guides, the pass rate plummeted to less than 2% across the nation. Technical schools that required the license for graduation were in a state of panic. Technicians who needed the license for obtaining or maintaining a job were frustrated, and many technical schools discontinued their FCC license preparation courses for lack of reliable resources.

The motivation behind this book is to provide an effective resource that can be used by the reader to prepare for the new FCC test. The effectiveness of this book speaks for itself. The first edition was released by the author in Woodburn, Oregon, in late June of 1987. Glenn Brusse, a senior electronics instructor at ITT Technical Institute in Boise, Idaho, quickly adopted it as a text for his FCC license preparation course. Ten of the eleven who passed the August 1987 FCC test in Portland, Oregon, were Glenn's students. This gave the Portland Field Office the distinction of having the highest pass rate in the nation, at 16%. The pass rate at the Portland office on the November 1987 test session climbed to 50%, largely as a result of this book and local FCC preparation seminars conducted by Mr. Brusse and the author. Mr. Brusse currently enjoys a 75% pass rate with his students. An increasing number of other technical schools have joined ITT Technical Institute in adopting the book as a text for FCC license preparation and avionics classes. Avionics shops, large firms, and the U.S. Coast Guard have regularly used this study guide.

In years past, the FCC offered First Class, Second Class and Third Class Radiotelephone licenses. The Third Class exam required the passing of Elements 1 and 2 (rules and regulations). The Second Class license exam required the passing of Elements 1 and 2 plus Element 3 (electronics theory). Finally, the First Class license exam required the passing of Elements 1, 2, 3 and Element 4 (advanced electronics). The commercial First Class and Second Class Radiotelephone licenses no longer exist. They have been replaced by one license called the General Radiotelephone Operator License. Components of Elements 1 and 2 (marine rules and regulations) have been integrated into this license exam. Additional marine, aviation, and international fixed public radio services regulations have been added. Finally, Element 4 has been totally eliminated.

The "revised" Element 3 exam consists of 100 multiple-choice questions. The passing grade for the examination is 75%, with 1% credit per question. Only after Element 3 is passed may the student take the Radar Endorsement exam (Element 8).

The General Radiotelephone Operator License qualifies the technician to maintain and repair certain communications equipment. For example, FCC regulations state that the General Radiotelephone Operator License is required in order to install, test or repair marine and aviation equipment. At other installations where FCC deregulation has taken place, the holder of this license is looked upon with respect. Some companies offer handsome salary increases to those who get this ticket. Many students, technicians, and engineers feel that having the license gives them a competitive edge in the job market.

# Introduction

The purpose of this book is to fill the gap that other study guides have left. The goal is to assist the student and prospective radiotelephone operator in passing the new General Radiotelephone Operator License exam. The book was written in a concise manner to enable the reader to learn the important information as quickly as possible, without including a lot of extraneous information. It is believed that one who wants to pass the FCC test appreciates relevant information essential to passing the test. For this reason, the book is not intended to be a complete coverage of the field of electronic communications. Rather, it is intended to be a complete coverage of information essential to passing the new FCC exam. There are already several excellent books that provide a thorough coverage of the field of communications. The reader is encouraged to refer to them for additional understanding and background where necessary.

To ensure success on the exam, it is important that the reader be familiar with all the information in this study guide. Important topics from the exam are presented in detail, emphasizing a thorough understanding. The chapters are developed around topics that actually appear on the FCC test. In some cases, some additional information is provided in an effort to prepare the reader for possible exam changes. Concepts have usually been stressed instead of rote memory of facts. If the reader understands a concept, it will be possible to answer any question on that subject—even if the figures are changed. It is easy for the FCC to change component values, supply voltages, and the wording of questions, so the reader must be prepared for this.

FCC field offices for obtaining information and an application for the exam are listed at the end of this Introduction.

The General Radiotelephone Operator License Exam is currently given twice per year, in February and August. The cost for taking the test is $35.00 and the deadline for making application is 30 to 45 days prior to the test, which is given during the first week of the testing month. The reader is strongly encouraged to contact the nearest FCC field office to get the latest details as soon as possible.

The rules and regulations are very important. Failure to study them will ensure failure on the exam. Many electronic engineers, and very competent technicians, have failed the exam for lack of knowledge of the rules and regulations. Study them well.

The glossary is placed early in the book to emphasize its importance. Although some topics are discussed in later chapters in greater detail, much important information contained in the glossary is not. Many exam topics are isolated and do not warrant a complete chapter in the book. In these cases, a full chapter would require the inclusion of much extraneous information not directly needed to pass the test. A thorough study of the glossary is essential to success on the FCC exam.

Study questions that follow the chapters are an important means of determining comprehension. Essay and completion answers have been chosen over multiple choice because they require greater recall. They expose the reader to correct answers instead of incorrect answers.

A chapter on Radar Fundamentals is included. It will help in preparing for the Element 8 (Radar Endorsement) exam. Readers who have used this book to pass the Radar Endorsement exam have told us that the Element 8 exam is much easier than the Element 3 exam. Because the radar exam may not be taken until after the Element 3 test is passed, the reader is encouraged to become thoroughly familiar with the information necessary for the Element 3 exam first. We have had reports from some people who studied for both Element 3, and Element 8, and came away from the test with no license because they failed Element 3. Please do not underestimate the difficulty of the Element 3 exam.

Exam-taking tips are provided. They should increase your potential for success with the exam.

Finally, FCC-type multiple choice exams have been provided as an indicator of your preparedness. It should be understood that these questions are not exact FCC test questions but are very similar to actual test questions. The values will differ and the wording of the questions will be more extensive on the FCC test. A thorough understanding of how the answers are calculated is vital.

This study guide contains sufficient information to give you the ability to pass the exam. However, a basic understanding of electronics is assumed. There are several excellent texts available that can provide the electronic fundamentals, if additional information is necessary.

For additional FCC exam preparation, I make available audio cassettes, a video program, and additional materials that supplement this book. FCC preparation seminars are also offered in major cities. For information, contact WPT Publications at 7015 NE 61st Ave, Vancouver, WA 98661 or call (206) 750-9933. Your success is important to us.

## Good luck

I say "good luck" with some hesitation because I found that the harder I study, the better is my luck. However, one successful student said "Hard study is only effective if you study the 'right information.'"

## FCC office addresses

ALASKA, Anchorage Office
Federal Communications Commission
6721 West Raspberry Rd.
Anchorage, AK 99502-1896
(907) 243-2153

CALIFORNIA, San Diego Office
Federal Communications Commission
4542 Ruffner St., Room 370
San Diego, CA 92111-2216
(619) 467-0549

CALIFORNIA, Los Angeles Office
Federal Communications Commission
Cerritos Corporate Tower
18000 Studebaker Rd., Room 660
Cerritos, CA 90701-3684
(213) 809-2096

CALIFORNIA, San Francisco Office
Federal Communications Commission
3777 Depot Rd., Room 420
Haywood, CA 94545-1914
(415) 732-9046

COLORADO, Denver Office
Federal Communications Commission
165 South Union Blvd., Suite 860
Lakewood, CO 80228-2213
(303) 969-6497/8

FLORIDA, Miami Office
Federal Communications Commission
Rochester Building, Room 310
8390 Northwest 53rd St.
Miami, FL 33166-4668
(305) 526-7420

FLORIDA, Tampa Office
Federal Communications Commission
2203 North Lois Ave., Room 1215
Tampa, FL 33607-2356
(813) 228-2872

GEORGIA, Atlanta Office
Federal Communications Commission
3575 Koger Blvd.
Koger Center-Gwinnett, Suite 320
Duluth, GA 30136-4958
(404) 279-4621

HAWAII, Honolulu Office
Federal Communications Commission
Waipio Access Rd.
P. O. Box 1030
Waipahu, HI 96797-1030
(808) 677-3318

ILLINOIS, Chicago Office
Federal Communications Commission
Park Ridge Office Center, Room 306
1550 Northwest Hwy.
Park Ridge, IL 60068-1460
(312) 353-0195

LOUISIANA, New Orleans Office
Federal Communications Commission
800 West Commerce Rd., Room 505
New Orleans, LA 70123-3333
(504) 589-2095

MARYLAND, Baltimore Office
Federal Communications Commission
1017 Federal Building
31 Hopkins Plaza
Baltimore, MD 21201-2802
(301) 962-2729

MASSACHUSETTS, Boston Office
Federal Communications Commission
NFPA Building
1 Batterymarch Park
Quincy, MA 02169-7495
(617) 770-4023

MICHIGAN, Detroit Office
Federal Communications Commission
24897 Hathaway St.
Farmington Hills, MI 48335-1552
(313) 226-6078

MINNESOTA, Saint Paul Office
Federal Communications Commission
693 Federal Building & U.S. Courthouse
316 North Robert St.
Saint Paul, MN 55101-1467
(612) 290-3819

MISSOURI, Kansas City Office
Federal Communications Commission
Brywood Office Tower, Room 320
8800 East 63rd St.
Kansas City, MO 64133-4895
(816) 926-5111

NEW YORK, Buffalo Office
Federal Communications Commission
1307 Federal Building
111 West Huron St.
Buffalo, NY 14202-2398
(716) 846-4511

NEW YORK, New York Office
Federal Communications Commission
201 Varick St.
New York, NY 10014-4870
(212) 620-3437

OREGON, Portland Office
Federal Communications Commission
1782 Federal Office Building
1220 Southwest 3rd Ave.
Portland, OR 97204-2898
(503) 326-4114

PENNSYLVANIA, Philadelphia Office
Federal Communications Commission
One Oxford Valley Office Building
2300 East Lincoln Hwy., Room 404
Langhorne, PA 19047-1859
(215) 752-1324

PUERTO RICO, San Juan Office
Federal Communications Commission
San Juan Field Office
747 Federal Building
Hato Rey, PR 00918-2251
(809) 766-5567

TEXAS, Dallas Office
Federal Communications Commission
9330 LBJ Expressway, Room 1170
Dallas, TX 75243-3429
(214) 767-5690

TEXAS, Houston Office
Federal Communications Commission
1225 North Loop West, Room 900
Houston, TX 77008-1775
(713) 861-6200

VIRGINIA, Norfolk Office
Federal Communications Commission
1200 Communications Cir.
Virginia Beach, VA 23455-3725
(804) 441-6472

WASHINGTON, Seattle Office
Federal Communications Commission
11410 NE 122nd Way, Suite 312
Kirkland, WA 98034
(206) 821-9037

# 1
# FCC rules and regulations

Although the main purpose of this summary of rules and regulations is preparation for the FCC (Federal Communications Commission) exam, do not underestimate the importance of this information. The bulk of the rules and regulations addresses the subjects of general operating procedures, emergency operating procedures, and technical standards of equipment. The knowledge of general operating procedures enables the radio operator to be effective and professional in manner. Good operating procedure leads to an efficient and noninterfering use of the shared radio spectrum. Knowledge of and adherence to the technical standards lead to the efficient transmission and reception of radio signals. Above all, a thorough understanding of emergency procedures can be a life and death matter. A radio operator involved in emergency communications said, ''There is no greater satisfaction than a person you helped later coming to you and telling you 'thanks.'''

## Glossary

The following is a summary of (FCC) rules and regulations. They are covered in detail in the Code of Federal Regulations, Title 47, Parts 1, 2, 13, 17, 23, 80, and 87, relating to the subject of telecommunications. This glossary is not intended to be a substitute for the complete Code of Federal Regulations. It is intended to serve as a quick reference to the more commonly used sections and to summarize the important sections that are covered on the FCC exam. The numbers in the parentheses correspond to the sections of the Code of Federal Regulations.

**Annual inspection**   (80.59) Compulsory ship stations must be inspected and certified annually. FCC Form 801 must be submitted to the Engineer in Charge of the FCC District Office nearest the proposed place of inspection. It must be sent at least THREE DAYS before the proposed inspection date. The same form may be used to apply for inspection of bridge-to-bridge radio inspection. FCC Form 808 must be used if the inspection is to be on a holiday or after normal working hours. A temporary waiver of annual inspection may be granted for a period not to exceed 30 days. Detailed inspection of compulsory radiotelephone installations of small passenger boats is required every five years. (80.903)

**Applications** (80.23, 1.921)

1. *New Ship station license* must be applied for at least 60 days before it will be needed. FCC Form 506 shall be used. It must be signed by:
   a. Vessel owner, or
   b. Vessel's operating agency, or
   c. Ship station licensee, or
   d. Master of the vessel
2. *Ship station license renewal* applications must be made during the license term and should be filed within 90 days but not later than 30 days prior to the end of the license term (1.926). The normal term of the station license is five years from the date of original issuance, major modification or renewal. FCC Form 405-B shall be used. If modification is needed, FCC Form 506 shall be used.
3. *Application for FCC ship radio inspection* must be submitted to nearest field office at least three days before desired inspection by one of the following (See Annual inspections):
   a. Vessel owner, or
   b. Vessel's operating agency, or
   c. Ship station licensee, or
   d. Master of the vessel

**Associated ship unit** (80.5) A portable VHF transmitter for use in the vicinity of the ship station with which it is associated. For more information, see section on *Portable ship units*.

**Authority of the master** (80.114) The service of each ship station must at all times be under the ultimate control of the master, who must require that each operator or such station comply with the radio regulations in force. These rules are waived when the vessel is under the control of the U.S. Government.

**Authorization of power** (80.63) In the interest of avoiding interference to other operations, all stations shall radiate only as much power as is necessary to ensure a satisfactory service. Designation of effective radiated power may appear on the station license. Except for transmitters using single sideband or independent sideband emissions, each radio transmitter rated for carrier power in excess of 100 W must contain instruments necessary to determine the transmitter power during operation.

**Auto alarms** (80.259, 80.261, 80.269, 80.317, 80.318, 80.811, 80.817). There are two types of auto alarms. The radiotelegraph auto alarm is operated on 500 kHz. The radiotelephone auto alarm is operated on 2182 kHz. You will note that these frequencies are specified for emergency traffic. (See Listening watch and Silent period.)

*Purpose* To attract attention of persons on watch or to activate automatic alarm devices. An auto alarm signal announces one of the following:

1. That a distress call is about to follow
2. That a transmission of an urgent cyclone warning is about to follow. This shall be used by coast stations
3. The loss of a person or persons overboard. The message must be preceded by the urgency signal

*Signal generator* The international radiotelephone alarm signal consists of two sinusoidal audio tones (2200 and 1300 hertz [Hz]) with a duration of 250 milliseconds (ms). May be transmitted from 30 to 60 seconds at a time.

*Testing requirements (keying device)* (80.811) The automatic radio telegraph alarm-signal keying device must be tested for correct timing adjustment of the keying mechanism. Do not transmit when making the test. Tests shall be made as follows:

1. Prior to the vessel's departure from each port
2. On each day that the vessel is outside of a harbor or port
3. If vessel is in two or more ports within one day, the required test need be made only once
4. If vessel is in port for less than one day, the required test for that day may be made before arrival or after departure

*Testing requirements (receiving unit)* (80.817) No reference to the testing of radiotelephone auto-alarm receiving equipment was found in the Code of Federal Regulations (CFR). However, the radiotelegraph auto-alarm testing was as follows: The radio officer must test the radiotelegraph auto alarm at least every 24 hours (h) while the ship is in the open sea as follows:

1. Test the auto alarm by using a testing device to determine whether it will respond to the proper tone sequence.
2. Determine the proper function of the auto alarm receiver by comparison of received signals on 500 kilohertz (kHz) by the main receiver.

**Bandwidth** (80.205) Authorized bandwidth is the maximum occupied bandwidth authorized to be used by a station. Bandwidths for various types of emissions are summarized in Table 1-1.

**Bandwidth of emission** (2.202) Occupied bandwidth is the frequency bandwidth where, below its lower and above its upper frequency limits, the mean powers radiated are equal to 0.5% of the total mean power radiated by a given emission.

1. Contains 99% of total radiated power
2. Contains carrier, sidebands, and harmonics
3. Includes any frequency containing 0.25% of the total radiated power

**Table 1-1. Bandwidths for various emission types.**

| | |
|---|---|
| A1A . . . . . . . . . . . 0.4 kHz | |
| A3E . . . . . . . . . . . . 8 kHz | |
| F3C . . . . . . . . . . . . 3 kHz | |
| F3C . . . . . . . . . . . 20 kHz | (Applicable only to facsimile in the 156 – 162 MHz and 216 – 220 MHz bands) |
| F3E . . . . . . . . . . . 20 kHz | (Applicable only when maximum frequency deviation is 5 kHz) |
| G3E . . . . . . . . . . . 20 kHz | (Applicable only when maximum frequency deviation is 5 kHz) |
| H3E . . . . . . . . . . . . 3 kHz | |
| J3E  . . . . . . . . . . . . 3 kHz | |
| R3E . . . . . . . . . . . . 3 kHz | |

**Bridge-to-bridge communications**   (80.163, 80.309, 80.331, 80.1001 – 1023)

*Purpose*   Navigational only.

*Applicability*   The following vessels must have bridge-to-bridge communication capability:

    a. Every power-driven vessel of 300 gross tons and upward while navigating

    b. Every vessel of 100 gross tons and upward carrying one or more passengers for hire while navigating

    c. Every vessel 26 ft (7.8 meters [m]) or over in length while navigating

    d. Every dredge and floating plant engaged, in or near a channel, in operations likely to restrict or affect the navigation of other vessels

*License requirement*   The operator must have a restricted radiotelephone operator permit or higher class license.

*Power*

- Generally is limited to 1 watt (W) or less.
- High power (more than 1 W) may be used *only* under the following conditions:
    1. In an emergency situation
    2. When rounding a bend in a river or navigating through a blind spot
    3. When a ship fails to respond to a call on low power

*Frequency*   Channel 13 (156.650 megahertz [MHz]) is the international bridge-to-bridge frequency. (63E emission)

*Who may operate bridge-to-bridge transmitters?*

    1. Master or person in charge of the vessel

    2. The person in charge of navigation of the vessel

*A continuous watch on channel 13 must be maintained*:

    1. By the master of the vessel or person in charge of navigation

    2. When vessel is within 100 miles (mi) of U.S. shores

*Station identification requirements*:
1. Name of vessel may be used, in lieu of the station call sign
2. ID at start and end of transmission and every 15 min during a long transmission

**Bridge-to-bridge station** (80.5, 80.1011) A radio station located on a ship's navigational bridge or main control station operating on a specified frequency (channel 13) which is used only for navigational communications, in the 156 – 162 MHz band.

**Calling frequency** (80.111) Initial contact may be made on a calling frequency. Calling a particular station must not continue for more than one minute in each instance. If there is no reply, the call may be sent three more times at intervals of two minutes. If there is still no reply, the station may be called again in 15 minutes (min). When contact is made, the two stations must move to a "working frequency" for the communications. To facilitate reception of distress calls, all transmissions on 2182 kHz and 156.8 MHz must not exceed one minute. Thus, the calling frequency is left clear for possible distress calls.

**Call signs** (2.302) Call signs are assigned by the Commission to various classes of stations. The call sign consists of a combination of letters and digits. For example, a ship radiotelephone call sign may consist of:
1. Two letters followed by four digits, as WA2000, or
2. Three letters followed by four digits, as WZZ9999

When stations operating in two or more classes are authorized to the same licensee for the same location, the Commission may elect to assign a separate call sign to each station in a different class.

**Carrier suppression** (See Emission types)

**Changes during license term** (80.29, 80.56) When changes are made during the license term, the Commission may have to be notified. Table 1-2 summarizes types of changes and required action.

It is important to know that only certain types of station licenses may be assigned. Section 1.924 of Title 47 contains the necessary information regarding this. This section states that licenses for stations in the Amateur, Aviation (aircraft), Personal, and Maritime (ship) Radio Services *cannot* be assigned. Whenever there is a change of ownership of one of these stations, the new owner must apply for a new license. Upon receipt of the new license, the former license must be surrendered for cancellation. The following may be assigned:
1. Coast stations in the Maritime Services and Alaska-Public-Fixed Stations may apply for assignment with FCC Form 503.
2. Ground stations in the Aviation Services may apply for assignment with FCC Form 406.

**Changes in authorized stations** (87.31, 87.35) An application for modification of license shall be filed when any change is to be made which would result in

### Table 1-2. Changes during license term.

| Type of change | Required action |
|---|---|
| Mailing address | Written notice to the Commission. |
| Name of licensee | Written notice to the Commission. |
| Name of vessel | Written notice to the Commission. |
| Addition of new transmitting equipment that operates on frequencies not authorized by present license | Application for modification. |
| Addition of transmitting equipment that operates on frequencies authorized by present license | None, provided the new equipment is type accepted and the emission characteristics remain the same. |
| Transfer of control of a corporation | Comply with Section 1.924 of this chapter (CFR Title 47) |
| Assignment of a radio station license | Comply with Section 1.924 of this chapter (CFR Title 47) |

deviation from the terms of the authorization. For example, anything that will affect the following:

1. Frequency tolerance
2. Modulation
3. Emission
4. Power
5. Bandwidth

No application for modification of radio station license is required for the following:

1. To add a survival craft station, using type accepted transmitters
2. For the addition or substitution of new transmitters, as long as they perform the same functions and operate on the same frequencies as the transmitters specified on the license

**Channel designations**   (80.373) Section 80.373 offers a detailed listing of the various frequency channels. The following is a list of several important frequency channels:

   a. Channel 6 (156.300 MHz) is used for intership safety and search and rescue communications (usually in conjunction with the U.S. Coast Guard)
   b. Channel 12 (156.600 MHz) is used for port operations, consisting of intership and ship-to-coast communications
   c. Channel 13 (156.650 MHz) is used exclusively for navigational bridge-to-bridge communications

    d. Channel 15 (156.750 MHz) is used for coast to ship environmental communications. It is also used in Class C EPIRBs (emergency position indicating radiobeacons) in conjunction with channel 16

    e. Channel 16 (156.800 MHz) is used for distress and safety calling

    f. Channel 22 (157.100 MHz) is used for communication with the U.S. Coast Guard. Contact with the Coast Guard can also be made on 2182 kHz, on the medium frequency band

**Class of emissions**   (See Emission types.)

**Classification of operator licenses**   (13.2) Commercial radio operator licenses and endorsements issued by the Commission are classified in accordance with the Radio Regulations of the International Telecommunications Union as follows:

1. First Class Radiotelegraph Operator's Certificate
2. Second Class Radiotelegraph Operator's Certificate
3. Third Class Radiotelegraph Operator's Certificate
4. General Radiotelephone Operator License
5. Marine Radio Operator Permit
6. Restricted Radiotelephone Operator Permit
7. Ship Radar endorsement
8. Six Months' Service endorsement (for First and Second Class Radiotelegraph Licenses)

The following former licenses and endorsements are no longer issued:

1. Radiotelephone First Class Operator License—last issued December 1981
2. Radiotelephone Second Class Operator License—last issued December 1981
3. Radiotelephone Third Class Operator License—last issued October 1980
4. Broadcast endorsement-last issued February 1979

**Compulsory ship**   (80.801, 80.851, 80.901) Any ship that is required to be equipped with radiotelecommunications equipment in order to comply with the radio-navigation provisions of a treaty to which the vessel is subject. There are compulsory radiotelegraph installations for cargo vessels of 1600 gross tons and upward, and for all passenger ships irrespective of size. Compulsory radiotelephone installations are required in cargo ships of 300 gross tons and upward but less than 1600 gross tons on international voyages. Radiotelephone installations are also required for small passenger boats that transport six or more passengers. By contrast, a "voluntary ship" is not required to be equipped with radiocommunication equipment.

**Contents of radio communications**   (80.88) It is unlawful to disclose the content of any radio communication to anyone except the party to whom the communication is addressed. The only exception is if you hear the message of a ship or aircraft in distress. An operator may not use any information heard, for his own benefit, or for the benefit of anyone else who is not entitled to receive it.

**Contents of station records**   (23.48, 80.409) Voluntary ships are not required to have a station log. The licensee of a compulsory ship shall maintain a technical log of the station operating showing:
1.  Signature of each licensed operator responsible for the operation of the transmitting equipment and an indication of his hours of duty
2.  Listing of the frequencies used, including:
    a.  Type of emission
    b.  Starting and ending times of each period of communications
    c.  Points of communications used
3.  Power input to the final stage of each transmitter
4.  Dates and results of frequency measurements

For stations in the *international fixed public control service*, the licensee shall maintain a technical log of the station operating showing:
1.  Normal hours of operation and dates and times of interruptions to service
2.  Dates and results of each frequency measurement
3.  When service or maintenance duties are performed, the responsible operator shall sign and date the station record giving details of all duties performed by him or under his supervision. He must also include his name and the class, serial number, and date of expiration of his license

**Control point (aviation stations)**   (87.75, 87.95, 87.103) The control point is the operating position of the station. Each station shall be provided with a "control point" at the location of the transmitting equipment. The control point must meet the following requirements:
1.  Such position must be under the control and supervision of the licensee
2.  It is the position where monitoring facilities are located
3.  It is the position where the transmitter can, without delay, be rendered inoperative
4.  It is the position at which the required licensed radio operator, responsible for the actual operation of the transmitter, is stationed
5.  The above does not apply to aeronautical enroute stations

*NOTE:* The current authorization, or a photocopy, for each station at a fixed location shall be prominently displayed at the principal control point of the transmitter, or transmitters.

**Control point (ship stations)**   (80.80) The control point is the operating position of the station. Each ship station control point must be capable of:
1.  Starting and discontinuing operation of the station
2.  Changing frequencies
3.  Changing from transmission to reception and vice versa
4.  Reducing transmitter power when necessary

*NOTE:* The current authorization, or a photocopy, for each station at a fixed location shall be prominently displayed at the principal control point of the transmitter, or transmitters.

**Cooperative use of frequency assignments** (80.87, 80.89, 80.92) Each radio channel is available for use on a shared basis only and is not available for the exclusive use of any one station or station licensee. Cooperation is important in order to have minimum interference.

*Station operators should*:

1. Listen to the frequency before starting the transmission. Make sure the frequency is not in use
2. Keep transmissions as short as possible

*Station operators should not*:

1. Engage in superfluous communications
2. Use selective calling on 2182 kHz or 156.8 MHz. (Selective calling is a method of transmission that allows reception only by specific stations. Tone-coded signals are used)
3. Transmit a general call or signals not addressed to a particular station
4. Transmit while on land

**Definitions (stations and services)** (2.1, 80.5, 87.5)

*Aeronautical mobile service* A mobile service between aeronautical stations and aircraft stations, or between aircraft stations, that survival craft stations may participate in.

*Aeronautical station* A land station in the aeronautical mobile service. In certain cases, the station may be located on board ship or on a platform at sea.

*Aircraft station* A mobile station in the aeronautical mobile service, other than a survival craft station, located on board an aircraft.

*Base station* A land station in the land mobile service carrying on a service with land mobile stations.

*Coast station* A land station in the maritime mobile service.

*Coordinated universal time (UTC)* UTC is equivalent to mean solar time at the prime meridian (0° longitude)—formerly called GMT (Greenwich Mean Time).

*Duplex operation* A mode of two-way communications where both persons may talk at the same time, like on the telephone. This requires the use of two frequencies: one for receiving and one for transmitting. (See Simplex.)

*Effective radiated power* Effective radiated power (e.r.p.) in a given direction is the product of the power supplied to the antenna and its gain relative to a half-wave dipole.

*Emergency position-indicating radiobeacon station* A station in the mobile service, the emissions of which are intended to facilitate search and rescue operations.

*Facsimile* A form of telegraphy for the transmission of fixed images. The images are reproduced in permanent form at the receiver.

*Fixed public service*   A radiocommunication service carried on between fixed stations open to public correspondence.

*Fixed service*   A service of radiocommunication between specified fixed points.

*Fixed station*   A station in the fixed service.

*Frequency tolerance*   The maximum permissible departure by the center frequency of the frequency band occupied by an emission from the assigned frequency.

*International fixed public radiocommunication service*   A fixed service, the stations of which are open to public correspondence and which, in general, is intended to provide radio communication between any one of the states or U.S. possessions or any foreign point, or between U.S. possessions and any other point. This service also involves the relaying of international traffic between stations that provide this service.

*Land mobile service*   A mobile service between base stations and land mobile stations, or between land mobile stations.

*Land mobile station*   A mobile station in the land mobile service capable of surface movement within the geographical limits of a country or continent.

*Land station*   A station in the mobile service not intended to be used while in motion.

*Maritime mobile service*   A mobile service between coast stations and slip stations, or between ship stations, or between associated on-board communication stations. Survival craft and EPIRB stations also participate in this service.

*Mobile service*   A service of radio communication between mobile and land stations, or between mobile stations.

*Mobile station*   A station in the mobile service intended to be used while in motion or during halts at unspecified points.

*Navigational communications*   Safety communications pertaining to the maneuvering of vessels or the directing of vessel movements. Such communications are primarily for the exchange of information between ship stations and, secondarily, between ship stations and coast stations.

*Operational fixed station*   A station that provides control, repeater or relay functions for its associated coast station.

*Point of communication*   This means a specific location designated in the license that a station is authorized to communicate for the transmission of public correspondence.

*RACON*   Radionavigation system.

*Radionavigation*   A system of determining position of a vessel by the use of radio waves.

*Survival craft station*   A mobile station in the maritime or aeronautical mobile service intended solely for survival purposes.

**Distortion**   (80.213) Distortion of a transmitter's signal can be caused by several internal malfunctions. However, one external way to distort a signal is by shouting into the microphone. Another way is by turning up the microphone gain control. Modulation beyond 100 percent can occur. This makes the signal less understandable rather than louder.

**Distress call** (80.311 – 80.332) Distress calls can be made only under the authority of the master of the vessel. The distress call has absolute priority over all other transmissions.

*Radiotelegraph distress procedure* A radiotelegraph distress signal is sent by International Morse Code at a speed of 8 – 16 words per minute. The procedure consists of the following:

1. Radiotelegraph alarm signal
2. The distress call, consisting of:
    a. The distress signal SOS, sent three times
    b. The word DE (meaning "THIS IS")
    c. The call sign of the station in distress, sent three times
3. The distress message, consisting of:
    a. The distress signal SOS
    b. The name of the station in distress
    c. Particulars of its position
    d. The nature of the distress
    e. The kind of assistance desired
    f. Any other information that might facilitate rescue
4. Two dashes of ten to fifteen seconds each
5. The call sign of the station in distress

*Radiotelephone distress procedure*

1. Radiotelephone alarm signal (whenever possible)
2. Distress call, consisting of:
    a. The distress signal MAYDAY, spoken three times
    b. The words THIS IS
    c. The call sign of the station in distress, spoken three times
3. The distress message, consisting of:
    a. The distress signal MAYDAY
    b. The name of the station in distress
    c. Particulars of the vessel's position
    d. The nature of the distress
    e. The kind of assistance required
    f. Any other information that may facilitate rescue (for example, the length, color, and type of vessel, number of persons on board, position of ship, etc.)
4. The distress transmission shall be made slowly and distinctly

**Distress frequencies** (80.313)

*Radiotelephone distress frequencies*

- 2182 kHz (in the medium frequency range). J3E emission is used.
- 156.8 MHz (channel 16, international distress and calling frequency) (in the VHF—very high frequency—range). F3E emission is used.

- 121.5 MHz (universal simplex clear channel for aircraft in distress). A3E emission is generally used.

*NOTE:* Except when making a distress call, it is illegal to transmit a general call on these frequencies, that is, a communication not addressed to a particular station.

*Radiotelegraph distress frequency*
- 500 kHz. A2B emission is used.

**Duplex**   A mode of operation in which two stations can talk at the same time. (See Simplex).

**Eligibility for new license**   (13.5, 80.15, 87.3) You are eligible for a commercial license if:

1. You are a U.S. citizen or an alien who is eligible for employment in the U.S
2. You hold a valid United States pilot certificate
3. You hold a foreign aircraft pilot certificate valid in the United States, provided there is a reciprocal agreement between the countries.

You are not eligible for a license if:

1. You are afflicted with complete deafness or complete muteness, or complete inability for any other reason to transmit correctly and to receive correctly by telephone spoken messages in English
2. Your license is under suspension, or is involved in suspension proceedings
3. You are involved in any pending litigation based on alleged violation of the Communications Act of 1934, as amended.

*NOTE:* No applicant who is eligible to apply for any commercial radio operator license shall, by reason of any physical handicap, other than those mentioned above, be denied the privilege of applying and being permitted to attempt to prove his qualifications (by examination if examination is required) for such operator license.

**Emergency locator transmitter (ELT)**   (87.5, 87.187, 87.193, 87.195) A transmitter intended to be actuated manually or automatically, and operated automatically as part of an aircraft or survival craft station, as a locating aid for survival purposes.

*Frequency*   The operating frequencies of the ELT are 121.500 MHz and 243.00 MHz, which are the same as used by Class A, Class B, and Class S EPIRBs. Emission types permitted include A3E, A3N, and NON.

*Operational tests*   ELTs may be tested, as follows:

1. Operator uses the manually activated test switch, which switches the transmitter's output to a dummy load
2. If the unit is not fitted with a manual test switch:
   a. Testing can be done in coordination with, or under the control of, an FAA (Federal Aviation Administration) representative to assure that no transmission of radiated energy occurs that will result in a false distress signal

b. If testing with FAA involvement is not feasible, brief operational tests are authorized. The tests are to be conducted within the first 5 minutes of any hour and must not be longer than three audio sweeps. A dummy antenna must be inserted if possible.

## Emergency position indicating radiobeacon's (EPIRB) (80.15, 80.833, 80.1051 – 80.1059)

*Purpose* The EPIRB is a station in the maritime mobile service the emissions of which are intended to facilitate search and rescue operations. The emergency locator transmitter (ELT) is the EPIRB's counterpart in the aviation service.

*General requirements*
1. Must be battery operated
2. Must be waterproof and weather resistant
3. Must have operating instructions permanently attached that are understandable to untrained personnel
4. The exterior must have no sharp edges
5. Must have a manually activated test switch

*Battery replacement* The month and year of the battery's manufacture must be permanently marked on the battery and the month and year upon which 50% of its useful life will have expired must be permanently marked on both the battery and the outside of the transmitter. The batteries must be replaced if 50% of their useful life has expired or if the transmitter has been used in an emergency situation.

The useful life of the battery is the length of time that the battery must be stored under marine environmental conditions without the EPIRB transmitter output power falling below 75 mW prior to 48 h of continuous operation.

*Operational tests*
1. Manual test switch is used
2. Antenna should be replaced with a dummy load to prevent radiation of a false distress signal
3. Tests may be coordinated with the U.S. Coast Guard
4. When Coast Guard involvement is not possible, brief tests are authorized, provided they are conducted during the first five minutes of any hour and are not longer than three audio sweeps or one second whichever is longer

### Classification of EPIRBs
There are several different classes of EPIRB. In addition to the information that will immediately follow, the reader is encouraged to study Table 1-3, which shows the similarities and differences of the classes.

### Class A EPIRB
*Authorization* Class A EPIRB's are required for vessels authorized to carry survival craft or for vessels expected to travel in waters beyond the range of marine VHF distress coverage. This is considered to be more

**Table 1-3. EPIRB summary.**

| | Class A | Class B | Class C | Class S |
|---|---|---|---|---|
| *Operating frequency* | 121.5 MHz 243.0 MHz | 121.5 MHz 243.0 MHz | 156.8 MHz 156.75 MHz | 121.5 MHz 243.0 MHz |
| *Effective radiated power* | 75 mW for 48 h | 75 mW for 48 h | Not less than 1 W | 75 mW for 48 h |
| *Emission type* | A3N | A3N | G3E with two-tone alarm | A3N |
| *Type of activation* | automatic | manual | manual | manual |
| *Antenna deployment* | automatic | manual | manual | manual |
| *Requirement* | More than 20 mi offshore | More than 20 mi offshore | Less than 20 mi offshore | On survival craft |

than 20 mi, or (32 km), off shore. In this case, a class A or a class B EPIRB must be used.

*Activation*   Activates automatically when it floats free of the sinking ship. The antenna automatically deploys when the EPIRB activates.

*Frequency*   121.5 MHz and 243 MHz (These two frequencies are used for Class A EPIRBs, Class B EPIRBs, Class S EPIRBs and ELT's.)

*Modulation*   A3N is mandatory. A3E and NON are optional. (These modulation types are used for Class A EPIRBs, Class B EPIRBs, Class S EPIRBs and ELTs.)

*Effective radiated power*   Must not be less than 75 mW after 48 h of continuous operation. (This power requirement is the same for Class A EPIRBs, Class B EPIRBs and Class S EPIRBs.)

### Class B EPIRB
*Authorization*   Same as for Class A EPIRBs.

*Activation*   By a water-activated battery or manually with a switch. Antenna is deployed manually.

*Frequency*   Same as Class A EPIRBs

*Modulation*   Same as Class A EPIRBs

*Effective radiated power*   Same as Class A EPIRBs

### Class C EPIRB
*Authorization*   For use on board vessels operating within 20 miles (32 km) offshore and in the Great Lakes or on passenger and cargo vessels with survival craft as required by the U.S. Coast Guard.

*Activation*   Manually with a switch.

*Frequency*   Dual frequencies are used: 156.750 MHz (ch 15) and 156.800 MHz (ch 16). Channel 16 is modulated first for 1.5 s. Then channel 15 is modulated for 14.5s, followed by another 1.5-s modulation of channel 16.

The remaining cycles are on channel 15.

*Modulation*   G3N emission is used, with a two-tone alarm signal, consisting of 2200 and 1300 Hz signals.

*Effective radiated power*   Must not be less than 1 watt.

### Class S EPIRB

*Authorization*   For survival crafts. Intended to be permanently attached to a survival craft. It is not required to float.

*Activation*   Manually with a switch.

*Frequency*   Same as Class A EPIRBs

*Modulation*   Same as Class A EPIRBs

*Effective radiated power*   Same as Class A EPIRBs

### 406 MHz satellite EPIRB

*Authorization*   May be used by any ship required by the U.S. Coast Guard or by any ship that is equipped with a VHF ship radio station.

*Activation*   Manually with a switch.

*Frequency*   406.025 MHz. with a "homing" beacon on 121.5 MHz.

*ID code*   Each unit must be registered with NOAA (National Oceanic and Atmospheric Administration). NOAA issues a unique ID code to the registered unit.

**Emission types**   (2.201, 80.207) Authorized classes of emission have been given standard designations. For example, A3E refers to double-sideband, full carrier amplitude modulated telephony. And J3E refers to single-sideband, suppressed carrier amplitude modulated telephony. A full explanation of how these designations can be found in chapter 3, *Emmission Types and Frequency Ranges*.

**Equipment and service tests**   (23.32)

1. *Equipment tests*   Upon completion of construction of a radio station in exact accordance with the terms of the construction permit, the permittee is authorized to test the equipment for a period not to exceed 10 days, provided that:
   a. The engineer in charge of the district in which the station is located is notified two days in advance of the beginning of tests
   b. The Commission reserves the right to cancel such tests, or suspend, or change the date of the tests if such action appears to be in the public interest, convenience, and necessity, by notifing the permittee
2. *Service tests*   When construction and equipment tests are completed in exact accordance with the terms of the construction permit, and after an application for station license has been filed with the Commission showing the transmitter to be in satisfactory operating condition, the permittee is authorized to conduct service tests for a period not to exceed 30 days, provided that:
   a. The engineer in charge of the district in which the station is located is notified two days in advance of the beginning of tests
   b. The Commission reserves the right to cancel such tests, or suspend, or

change the date of the tests if such action appears to be in the public interest, convenience, and necessity, by notifying the permittee

c. Service tests will not be authorized after the expiration date of the construction permit

*NOTE:* The authorization for the above tests shall not be construed as constituting a license to operate but as a necessary part of the construction.

**Frequency deviation**   (80.213) Transmitters using F3E or G3E emission in the band 156 – 162 MHz or G3E in the band 218 – 220 MHz, shall be capable of proper technical operation with a FREQUENCY DEVIATION OF ±5 kHz, which is defined as 100% modulation.

**Frequency measurements (Aviation)**   (87.71, 87.111) Measurements of the operating frequencies of airborne transmitters may be required by the Commission in individual circumstances. The operating frequencies of all *non airborne* transmitters authorized for operation in the Aviation Services shall be measured at the following times to assure compliance with the tolerances specified in these rules:

1. When a transmitter is originally installed
2. When any change or adjustment is made in a transmitter that may affect an operating frequency
3. Whenever there is reason to believe that an operating frequency has shifted beyond the applicable tolerance
4. Upon receipt of an official notice of off-frequency operation

**Frequency measurements (Marine)**   (83.111 from CFR, October, 1985 edition) A determination shall be made that the carrier frequencies of each transmitter are within prescribed tolerance as follows:

1. When a transmitter is originally installed
2. When any change or adjustment is made in the transmitter that may affect the carrier frequency or stability thereof
3. Upon receipt of an official notice of off-frequency operation

**Frequency tolerances for transmitters**   (80.209, 87.133) The carrier frequency of transmitters must be maintained within acceptable tolerances. The tolerance is sometimes specified in parts per million (ppm) and other times specified in hertz. In many cases, the tolerance varies on different frequency bands. And the tolerance also depends on when the transmitter was type accepted. For a complete listing of specific tolerances, see the above rules and regulations references.

Ship transmitters are allowed to deviate a maximum of 20 Hz for frequencies below 156 MHz. Above 156 MHz, the tolerance drops to 10 Hz, for transmitters using 3 W or less.

All aircraft stations other than Civil Air Patrol have a tolerance of 20 Hz. Civil Air Patrol stations are required to stay within 50 Hz of tolerance. All aeronautical stations on land other than Civil Air Patrol are required to stay within 10 Hz. Table 1-4 summarizes several important frequency tolerances.

**Table 1-4. Frequency tolerance.**

Aeronautical stations on land . . . . . . . 10 Hz
Aircraft stations . . . . . . . . . . . . . . . . . 20 Hz
Civil Air Patrol stations . . . . . . . . . . . 50 Hz
Ship stations (up to 156 MHz) . . . . . . 20 Hz
Ship stations (above 156 MHz) . . . . . . 10 Hz
Survival craft stations . . . . . . . . . . . . . 50 Hz

**General radiotelephone operator license**   (13.2, 80.155, 80.159, 80.169, 87.133, 87.135) This license conveys all the operating authority of the Marine Radio Operator Permit. No person may therefore hold both at the same time. All adjustments of radiotelephone transmitting equipment in any *maritime coast or ship station*, made during installation, servicing, or maintenance of that equipment, which may affect its proper operation, may be made only by, or under the immediate supervision of, a person who holds a General Radiotelephone Operator License, or a First, or Second Class Radiotelephone Operator's Certificate.

Adjustments or tests relating to the installation, servicing or maintenance of an *aviation radio station* that might affect its proper operation, may be made only by, or under, the immediate supervision of a person who holds a General Radiotelephone Operator License, or a First or Second Class Radiotelephone Operator's Certificate.

**Inspection by foreign governments**   (80.79) The radio operator on board a ship must allow officials of foreign governments to examine the radio station license if they so request. The radio operator must facilitate the examination in order that the officials may be satisfied that the station complies with international radio regulations.

**License term**   (80.25, 87.51) Ship stations are licensed for a period of five years. Licenses for stations in the Aviation Services will nominally be issued for a term of five years from the date of original issuance, major modification, or renewal. First and Second Class Licenses were formerly issued for five years. The General Radiotelephone Operator License is now issued for life.

**Listening watch**   (80.146 – 80.148, 80.304 – 80.310) All ship and coast stations licensed to transmit radiotelephony are required to maintain an efficient and continuous watch on the RADIOTELEPHONE DISTRESS FREQUENCY (2182 kHz) whenever such station is not being used for communication. The watch may be conducted from:
1. The principal operating position, or
2. The room from which the vessel is normally steered
3. The listening watch must be maintained for at least three minutes two times each hour beginning on the hour and on the half hour. All stations

must refrain from transmitting on that frequency at those times (silent period) except for distress, urgency or safety messages).

*Coast stations* operating in the 156–162 MHz band and serving in rivers and inland lakes except the Great Lakes must maintain a safety watch on 156.8 MHz.

*Telegraphy stations* must maintain a silent period on 500 kHz of three minutes twice each hour beginning at 15 minutes and 45 minutes after each hour. Do not confuse this with the radiotelephone silent period outlined above.

**Logs**   (23.48) For stations in the *international fixed public control service.*, the licensee shall maintain a technical log of the operating station showing:
1. Normal hours of operation and times of interruptions to service
2. Dates and results of each frequency measurement
3. When service or maintenance duties are performed, the responsible operator shall sign and date the station record giving pertinent details of all duties performed by him or under his supervision

**Logs**   (23.47, 23.48, 80.409, 80.413, 80.1153) *Ship radiotelephone stations* shall maintain a log. Pages of the log shall be numbered in sequence and each page shall include the name of the vessel, the call sign of the station, country of registry, and the official number of the vessel. The station licensee and the radio operator in charge of the station are responsible for the maintenance of station logs. The following entries are required:
1. A summary of all distress, urgency, and safety traffic
2. A summary of communications conducted on other than VHF frequencies
3. A reference to important service incidents
4. The position of the ship at least once per day
5. The name of the operator at the beginning and end of the watch period
6. The time the watch begins when the vessel leaves port, and the time it ends when the ship reaches port
7. The time the watch is discontinued and the reason
8. The time when storage batteries are charged
9. The results of required equipment tests
10. The results of inspections and tests of compulsorily fitted lifeboat radio equipment
11. A daily statement about the condition of the required radiotelephone equipment, as determined by communications test
12. When the master is notified about improperly operating radiotelephone equipment
13. Pertinent details of all installation, service, or maintenance work done. The technician shall log in all the pertinent work he himself has done and must sign his name.

*Retention of logs*   Station logs shall be retained by the licensee as follows:
1. Routine logs must be maintained for a period of one year from the date of entry

2. Logs involving distress communications shall be retained for a period of three years from the date of entry
3. Logs that include communications that are under FCC investigation shall be retained until such time as the Commission specifically authorizes them to be destroyed

**Lost license**   (13.71) If your license is lost or destroyed, you must apply for a new one. Application must be made, with an explanation of how it was lost. If the lost license is later found, it (or the duplicate) must be returned to the FCC for cancellation.

**Main transmitter**   (80.253) The main transmitter must be equipped to measure:
1. Antenna current
2. Transmitter power supply voltages
3. Anode or collector currents
   Antenna power must be determined at the operating carrier frequency by the product of the antenna resistance and the square of the average antenna current ($P = I^2 R$), both measured at the same point in the antenna circuit.

**Maintenance tests of licensed stations**   (23.43,80.96,87.93) Licensed stations are authorized to carry out routine tests as may be required for proper maintenance of the stations. The tests must not cause interference with the service of other stations.

**Marine radio operator permit**   (13.4, 13.28, 80.167) A permit issued by the FCC for a period of five years. A person may not hold this permit and a GROL (General Radiotelephone Operator's License) at the same time. The holder of this permit is allowed to operate only those transmitters that:
1. Have simple, external switching devices
2. Do not require manual adjustment of frequency-determining elements

**Modulation requirements**   (80.213)
1. When double-sideband emission is used, the peak modulation must be maintained between 75 and 100%
2. When frequency or phase modulation is used (in the 156 – 220 MHz bands), the peak modulation must be maintained between 75 and 100%
3. Transmitters using A3E, F3E, and G3E emission must have a modulation limiter to prevent any modulation over 100%
4. Ship stations using G3D or G3E emission (in the 156 – 220 MHz bands) must be capable of maintaining proper frequency deviation of ±5 kHz

**Nameplate**   (80.1021) A durable nameplate shall be mounted on the required radiotelephone transmitting and receiving equipment. The nameplate shall show at least the name of the manufacturer and the type or model number.

**Navigational communications**   (80.5, 80.1007, 80.1011) Safety communications pertaining to the maneuvering of vessels or the directing of vessel movements. Such communications are primarily for the exchange of information between ship stations or between ship stations and coast stations. (See Bridge-to-bridge communications.)

**Need for licensed commercial radio operators**   See General Radiotelephone Operator License and Marine Radio Operator Permit.

**Nomenclature of frequencies**   (R & R 2.101) This is the classification of frequency ranges. For example, VHF (very high frequency) is the band of frequencies from 30 to 300 MHz. This is covered in detail in chapter 3 *Emission Types and Frequency Ranges*. An easy method of learning these frequency ranges is provided in that chapter.

**On-board communications**   (80.1175) A low-powered mobile station in the maritime mobile service. On-board stations communicate with:
1. Other units of the same station for operational communications
2. On-board stations of another ship or shore facility to aid in off pollution prevention during transfer of 250 or more barrels of oil
3. Other units of the same station in the immediate vicinity of the ship for operational communications related to docking, life boat emergency drills or in the maneuvering of cargo barges and fighters

**Operation of ship transmitter**   (80.89) The ship station may not be operated from any location except the ship. It is illegal to operate it while the ship is being transported, stored, or parked on land.

**Operating procedure**   (FCC Bulletin FO-32)
*To initiate a call to another ship*
1. Monitor the frequency before transmitting. Make sure it is clear. This is one of the most important things for a radio operator to learn
2. Initiate the call on the calling frequency. This is used only for making initial contact. Transmission should not exceed one minute on 2182 kHz or 156.8 MHz. (These should be left clear for possible emergency use.)
3. Move to a working frequency for the communication

*Procedure words*
1. "Over" means you expect a reply from the station you are talking with
2. "Clear" or "out" means your communication is complete and you do not expect a reply
3. "Roger" means that you received the transmission correctly

*Restrictions*   You may not engage in any of the following:
1. Unnecessary transmissions
2. Unidentified transmissions
3. Superfluous communications
4. Idle chit-chat
5. Communications containing obscene, indecent, or profane words or meaning.
6. Transmission of false call signs
7. Willful or malicious interference
8. Signals not addressed to a particular station

**Operator's responsibility** (13.63) The licensed operator responsible for the maintenance of a transmitter may permit other persons to adjust a transmitter in his presence for the purpose of carrying out tests or making adjustments requiring specialized knowledge or skill, provided that he shall not be relieved thereby from responsibility for the proper operation of the equipment.

**Other forms of station identification** (2.303) The following table indicates forms of identification that may be used in lieu of call signs by the specified classes of stations:

1. *Aircraft (U.S. registry)* telephone   Registration number preceded by the type of the aircraft, or the radiotelephony designator of the aircraft operating agency followed by the flight identification number
2. *Aircraft (foreign registry)* telephone   Foreign registry identification consisting of five characters. This may be preceded by the radiotelephony designator of the aircraft operating agency or it may be preceded by the type of the aircraft
3. *Aeronautical*   Name of the city, area, or airdrome served, together with such additional identification as may be required
4. *Aircraft survival craft*   Appropriate reference to parent aircraft, for example, the air carrier parent aircraft flight number or identification, the aircraft registration number, the name of the aircraft manufacturer, the name of the aircraft owner, or any other pertinent information
5. *Public Coast* (radiotelephone) and *Limited Coast* (radiotelegraph)   The approximate geographic location in a format approved by the Commission
6. *Fixed*   Geographic location
7. *Land mobile (public safety, forestry conservation, highway maintenance, local government, shipyard, land transportation and aviation services)*   Name of station licensee (in abbreviated form if practical), or location of station, or name of city, area, or facility served. Individual stations may be identified by additional digits following the more general identification.
8. *Land mobile (railroad radio service)*   Name of railroad, train number, caboose number, engine number, or name of fixed wayside station. A railroad's abbreviated name or initial letters may be used. Unit designators may be used in addition to the station identification to identify an individual unit or transmitter of a base station.

**Passing score** (13.24) To pass a written examination, an applicant must answer at least 75% of the questions correctly.

**Percent modulation** (80.213, 87.73) In general, marine transmitters shall be adjusted so that transmission of speech produces peak modulation percentages between 75 and 100%. Aviation regulations state between 70 and 100%.

**Phonetic alphabet** (FCC Bulletin FO-32) An internationally recognized alphabet. It may be used when communications are difficult or when signals are weak. Words are simply spelled out using the standard phonetic alphabet. For example, "RADIO" would be spelled: "ROMEO, ALFA, DELTA, INDIA, OSCAR."

**Plurality of stations** (80.37, 80.115) One station license may be issued to authorize a designated maximum number of marine utility stations operating at temporary unspecified locations.

**Points of communication** (80.505) Private coast stations and marine utility stations are authorized to communicate:
1. With any mobile station for the exchange of communications
2. With any land station for the purpose of aiding the exchange of safety communications
3. With ship stations

A private coast station and associated marine utility stations serving and located on a shipyard regularly engaged in construction or repair are authorized to communicate between stations when they are licensed to the same entity and communications are limited to serving the needs of ships on a noninterference basis.

**Portable ship units** (Also called hand helds, walkie-talkies, and associated ship units) (80.115)

*License* When at sea, they are under the authorization of the ship station license

*Power* Limited to 1 W.

*Frequency* Must be equipped to transmit on:
1. Channel 16 (156.800 MHz) (safety and calling frequency)
2. And at least one appropriate intership frequency

*May be used*
1. Only to communicate with the ship station with which it is associated or with associated ship units of the same ship
2. For operational communications aboard a ship

*May not be used*
1. From shore
2. To communicate with other vessels not associated with their vessel

*Station identification requirements*
1. Operator must give the call sign of vessel with which it is associated, followed by an appropriate unit designator
2. Must be done at least every 15 min.

*NOTE:* These must not be confused with portable units licensed as marine utility stations. These may be operated either on a ship or on shore.

**Posting of operator license** (80.407) The original authorization of each operator must be posted at the principal control point of the station. In lieu of posting, an operator who holds a restricted radiotelephone operator permit or a higher class operator license may have the operator authorization or a photocopy thereof available for inspection upon request by commission employees.

**Posting of station license**   (80.403, 80.405, 87.97) The current station authorization or a clearly legible copy must be posted at the principal control point of each station. If a copy is posted, it must indicate the location of the original. When the station license cannot be posted, as in the case of a marine utility station operating at temporary unspecified locations, it must be kept where it will be readily available for inspection.

**Power**   (80.215) The power that may be authorized in the Aviation Services shall not be greater than the minimum required for satisfactory technical operation. It shall be specified in the following terms:

1. Amplitude modulation transmitters  For stations using amplitude modulation with both sidebands and full carrier, authorized power shall be specified in terms of the unmodulated radio frequency carrier power at the transmitter output terminals.
2. Other transmitters  For stations using other types of modulation, power authorized shall be specified in terms of peak envelope power at the transmitter output terminals. PEP (peak envelope power) is defined as the mean power during one radio frequency cycle at the highest crest of the modulation envelope.

Power can be determined either by direct measurement or by multiplying the plate input power to the final amplifier by an appropriate factor.

**Priority of communications**   (80.91) The various categories of communications have been arranged by order of priority as follows:

1. Distress calls
   a. Indicate that a mobile station is threatened by grave and imminent danger and request immediate assistance
   b. MAYDAY is the radio*telephone* distress signal
   c. SOS is the radio*telegraphy* distress signal
2. Urgency signals
   a. Indicate that the calling station has a very urgent message concerning the safety of the ship, aircraft, or other vessel, or the safety of a person
   b. In radiotelephony, the signal consists of the word PAN, spoken three times
   c. Shall be sent only on the authority of the master of the ship or person responsible for the mobile station
   d. Mobile stations that hear the urgency signal shall continue to listen for at least three minutes. At the end of this period, if no urgency message has been heard, they may resume their normal service.
3. Safety signals
   a. Indicate that the station is about to transmit a message concerning the safety of navigation or give important meteorological warnings
   b. In radiotelephony, the signal consists of the word SECURITY, spoken three times

  c. All stations hearing the safety signal shall listen to the safety message until they are satisfied that the message is of no concern to them. They shall not make any transmission likely to interfere with the message.

 4. Radio direction-finding communications

 5. Navigation communications

  a. Communications relating to the navigation and safe movement of aircraft engaged in search and rescue operations.

 6. Navigation communications

  a. Communications relating to the navigation, movements and needs of ships, and weather observation messages destined for an official meteorological service.

 7. Government communications

 8. Service communications

 9. All other communications

**Secrecy of communication**  (80.88) Secrecy must be observed with all radio communications transmitted or received. The content must not be given to anyone or used for personal profit. (See Violations.)

**Shipboard antennas**  (80.863, 80.876) Marine antennas within the frequency band 30 – 200 MHz shall be vertically polarized. Lower frequency antennas may be horizontally polarized and as nondirectional as possible.

**Silent period**  (80.301, 80.304)

 1. Shall be observed at each hour and each half hour for a period of 3 minutes. For example, from 10:00 to 10:03 and 10:30 to 10:33. See Fig. 1-1.

 2. Except for messages of distress, urgency, and vital navigational warnings, ship stations shall not transmit on 2182 kHz during the silent periods.

*NOTE:* There is a similar silent period on 500 kHz for radiotelegraph stations. It takes place 15 min later than the radiotelephone silent period. (See Listening watch.)

**Simplex**  Two-way communications where both stations operate on the same frequency. When operating in the simplex mode, you must wait until the other station stops transmitting before you start transmitting. When using simplex operation, the operator should pause for a moment before transmitting in case someone needs to break in with an emergency, or to make contact with another station. (See Duplex.)

**Single sideband**  A type of emission where the one sideband is filtered out. The carrier can be reduced or suppressed. When single sideband is used, the UPPER sideband (higher frequency) is selected. (See Emission types.)

**Squelch control**  A control on the radio receiver that tunes out the background noise when the receiver is not receiving communications. Set the control just beyond the point where the background noise is cut off. When received signals

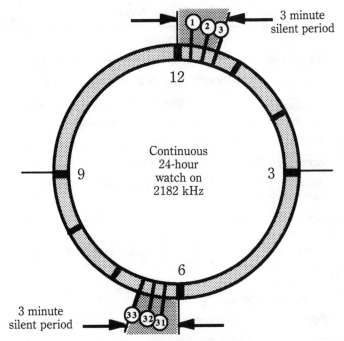

**1-1** Continuous watch and silent periods.

are very weak they may not be much stronger than the background noise level. If this is the case, it may be necessary to "open up" the squelch control so that the station can be heard.

**Standard forms** (1.922, 80.19, 87.29) Forms can be obtained from the Federal Communications Commission at Gettysburg, Pennsylvania, 17325; Washington, DC 20554, or any District Office. Table 1-5 lists many of the more commonly used FCC forms.

**Station identification** (2.303, 80.102, 80.104, 80.331, 87.115)
    *Aircraft stations*
      1. Aircraft radio station call sign, or
      2. Abbreviated call sign consisting of:
        a. Type of aircraft, followed by
        b. The last three characters of the registration marking, or
      3. Assigned FCC control number (for ultralight aircraft), or
      4. The radiotelephony designator of the aircraft operating agency (assigned by the FAA) followed by the flight identification number (assigned by the company)
    *Bridge-to-bridge stations*
      1. Name of vessel may be used in lieu of the station call sign
      2. Identification shall be made at the beginning, and upon completion of each transmission. The intervals shall not exceed 15 min

**Table 1-5. FCC forms.**

| FCC Form | Title |
| --- | --- |
| 404 | Application for new or modified aircraft radio station license. |
| 404-A | Application for temporary Aircraft Radio Station Operating Authority |
| 405-A | Application for renewal of radio license without modification. |
| 405-B | Application for license renewal without modification. |
| 406 | Application for Ground Station Authorization in the Aviation Services. |
| 480 | Application for Civil Air Patrol Radio Station Authorization. |
| 503 | Application for Land Radio Station License (or modification) in the Maritime Services. |
| 506 | Application for renewal of Ship Radio Station License. Also used in application for modification of license. |
| 506-A | Temporary operating authority, Ship Radio Station License and Restricted Radiotelephone Operator Permit. |
| 525 | Application for Disaster Communications Radio Station Construction Permit and License. |
| 801 | Application for Radio Inspection and Certification. |

*Land station in the aviation service*
1. Station call letters, or
2. Its location, or
3. Assigned FAA identifier, or
4. Name of the city

*Portable ship units*
1. Operator must give call sign with which he is associated, followed by a unit number, designating which mobile unit he is using.
2. Identification shall be made at the beginning, and upon completion of each transmission. The intervals shall not exceed 15 min.

*Private coast stations*   Stations located at drawbridges and transmitting on channel 13 may identify by use of the name of the bridge in lieu of the call sign.

*Radar transmitters*   Must NOT transmit station identification.

*Ship stations*
1. Ship stations using radiotelephony shall identify by announcing in the English language THE STATION ASSIGNED CALL SIGN
2. An exception to the above is where bridge-to-bridge channel 13 (156.65 MHz) is used. In this case, identification may be made with the NAME OF THE SHIP in lieu of the call sign
3. When the official call sign is not assigned by the Commission, the name of the ship on which the station is located or the name of the licensee may be used

    4. Identification shall be made at the beginning and upon completion of each transmission. The intervals shall not exceed 15 min.

*Survival craft stations*

1. The call sign of parent aircraft, or
2. Official aircraft registration of parent aircraft, followed by a single digit other than 0 or 1
3. No identification is required when distress signals are transmitted automatically

**Station license**   (80.13, 80.56) It is not legal to operate a radio station within the United States or its territories unless it is properly licensed in accordance with the Communications Act of 1934. Willful or repeated operation of an unlicensed radio station is punishable by fines or imprisonment. (See Violations.) Licenses may not be assigned or transferred when ownership of the vessel is transferred. The new owner must file for a new authorization.

**Survival craft station**   (80.5, 80.829-835) A mobile station in the maritime or aeronautical mobile service is intended solely for survival purposes, and is located on any lifeboat, life raft or other survival equipment. General requirements for survival craft radio equipment are as follows:

1. Must be buoyant
2. Equipment must be operated without tools
3. Must have instruction manual for maintenance
4. Simple instructions must be prominently attached
5. Transmitter must be pretuned to the required frequencies
6. It must be able to transmit the international radiotelegraph distress signal, alternately on 500 kHz and 8364 kHz
7. Radiotelephone transceivers must be able to transmit and receive on either 457.525 MHz or on 156.8 MHz. Power must be at least 0.1 W.

**Suspension of license**   (1.85) When grounds exist for suspension of an operator license, the suspension shall not take place until 15 days after the notification of the operator. The operator must make written application to the Commission (a hearing) within that 15-day period. The license must be sent to the Commission in Washington.

**Suspension of transmission**   (5.153, 80.90) If any deviation from the technical requirements is detected, transmission shall be immediately suspended until such malfunction is corrected. Examples of such deviation are:

1. If the transmitter is off frequency
2. If the transmitter becomes distorted
3. If the modulation exceeds 100%.

The only exception shall be in the event of an emergency, where temporary communications can be resumed until the emergency condition is over. At that time, transmission shall again be terminated for necessary repair work.

**Temporary operating authority**   (1.925, 87.41) An applicant for a new aircraft radio station license may operate the aircraft radio station pending issuance of a station license for a period of 90 days under a temporary operating authority. This is evidenced by a properly executed certification made on FCC Form 404-A. The applicant shall use the aircraft FAA registration number as a temporary call sign. Form 506-A is used for temporary operating authorities on a ship.

**Test of radiotelephone installation**    (80.1023) Unless normal use of the required radiotelephone installation demonstrates that the equipment is in proper operating condition, a test communication for this purpose must be made by a qualified operator each day the vessel is navigated. If the equipment is not in proper operating condition, the master must be notified promptly. The master must have it restored to effective operating condition as soon as possible.

**Testing procedure**    (80.101) Station licensees must not cause harmful interference. When radiation is necessary and unavoidable, the following testing procedure must be followed:
1. Operator must not interfere with transmissions in progress
2. The testing station's call sign, followed by the word TEST must be announced
3. If any station responds WAIT, the test must be suspended for a minimum of 30 s.
4. Tests must not exceed 10 s, and must not be repeated until at least 1 min has elapsed. On 2182 kHz or 156.0 MHz, the time between tests must be a minimum of five minutes
5. Operator must end the test with the station's call sign

**Tower lights**    (23.39, 87.113) If aviation tower lights are not monitored by automatic means, they must be inspected every 24 h. Appropriate entries shall be made in the station's technical log, as required by Section 23.39. The lights must operate from sunset to sunrise, and must be checked daily. Any observed or known failure of a top light or rotating beacon light not corrected within thirty minutes shall be reported immediately to the nearest Flight Service Station or office of the Federal Aviation Administration. Part 17 of Title 47 covers this in more detail. All flashing or rotating beacons and automatic lighting control devices must be inspected at least once every 3 months to ensure such apparatus is functioning properly.

**Transfer of station license**    (80.56, 87.33) Transfer or assignment is prohibited. Whenever a ship or aircraft is sold, the previous license must be cancelled. The new owner must file for a new license.

**Transmitter adjustments and tests by operator**    (87.135) All transmitter adjustments or tests during or coincident with the installation, servicing, or maintenance of a radio station, that may affect the proper operation of such station, shall be made by or under the immediate supervision and responsibility of a person holding a radiotelephone or radiotelegraph First Class or Second Class Operator License who shall be responsible for the proper functioning of the station equipment.

**Transmitter measurements**   (87.63, 87.73, 87.111)
1. Carrier frequency shall be checked as follows:
   a. During the initial installation
   b. When any change is made in the transmitter that may influence the operating frequency
   c. Upon receipt of an official notice of off-frequency operation.

*NOTE:* The output of the transmitter shall be checked against a suitable frequency standard. The National Bureau of Standards sets the frequency standard. A transmitter using crystal control shall have its frequency checked at least once per year.

2. Modulation Percent.The transmitter modulation shall be properly adjusted to produce peak modulation between 75 and 100%
3. Power Measurements. When the manufacturer's rated power of a transmitter is greater than 120% of the authorized power, the actual power output of the transmitter shall be determined as follows:
   a. During the initial installation
   b. When any change is made in the transmitter that may influence the power output.

*NOTE:* The carrier power of a ship station radiotelephone transmitter must be at least 8 W but not more than 25 W. All transmitters must be capable of reducing the carrier power to 1 W or less.

**Transmitter 150-m range requirement**   (80.807) The radiotelephone transmitter, in all radiotelegraph equipped vessels, must be capable of transmission of A3E or H3E emission, on 2182 kHz, clearly perceptible signals from ship to ship during daytime under normal conditions and circumstances over a minimum range of 150 nautical miles.

**Type acceptance**   (2.901-908, 2.961-969, 2.981-1003) Equipment that meets FCC specifications is "Type accepted." The "Radio Equipment List, Equipment Acceptable for Licensing" is available from any FCC District Office.
1. Transmitters may not be modified in their basic design in any way that would affect their output power, operating frequency, frequency stability or modulation percentage
2. Type accepted transmitters will have the following on the transmitter name plate
   a. Name of manufacturer
   b. FCC ID number

*Type approval*   The FCC conducts tests on the equipment to assure that it meets specifications.

*Type acceptance*   The manufacturer tests the equipment to assure that it meets specifications.

**Violations**   (5.162, 13.62, 80.149, FCC Bulletin FO-32) No licensed commercial radio operator shall violate or cause, aid, or abet the violation of any Act, treaty, or convention binding on the United States, which the Commission is authorized to administer, or any regulation made by the Commission under such Act, treaty, or convention.

1. If a person receives a notice of violation, a written response must be made with 10 days. It shall contain a full explanation of the incident and shall set forth the action taken to correct the deficiency and prevent its recurrence
2. Fines and penalties for violations are summarized in Table 1-6

**Table 1-6. Violations and penalties.**

| Violation | Not more than | Not more than | Or both |
|---|---|---|---|
| Failure to comply with any provisions of the Communications Act or FCC Rules and Regulations. | $2,000 for each violation | | |
| Willful violation of the Communications Act. | $1,000 | 6 months imprisonment | X |
| Willful violation of the Communications Act for commercial advantage or private financial gain. | $25,000 | 1 year | X |

**Voluntary ship** (80.1153) Licensees of voluntary ships are not required to operate the ship radio station or to maintain radio station logs. (See Compulsory ship)

When a ship radio station of a voluntary ship is being operated, appropriate general purpose watches must be maintained. (See Listening watch)

**Watch** See Listening watch.

**Working frequency** After establishing communications with another station by call and reply on 2182 kHz or 156.800 MHz (Calling Frequencies) you must change to an authorized working frequency for the transmission of the messages.

# Study questions
# Rules and regulations

The following questions highlight many important points. However, make an effort to become very familiar with the entire glossary of rules and regulations.

1. How many days in advance must you submit the FCC Form 801 for a transmitter inspection?
2. Who must apply for the above inspection?
3. If you need a new ship station license 6 months from now, when must you apply for it?
4. What is an *associated ship unit*?

5. Who has the highest authority on a ship?
6. What is the purpose of the auto alarm system?
7. What three things could an auto alarm be announcing?
8. What is the bandwidth of A3E emission?
9. Define *bandwidth of emission*.
10. For what purpose *bridge-to-bridge* communications?
11. Bridge-to-bridge communications are usually limited to 1 W of transmitter power. Under what three conditions may this power be increased?
12. What channel is used in bridge-to-bridge communications?
13. What is a calling frequency?
14. When a ship is sold, can the station license be assigned to the new owner?
15. Under what conditions must you apply for modification of your station license?
16. What is a compulsory ship?
17. What information must be contained in the station records?
18. When a technician performs service or repairs, what must the technician write in the log?
19. What does *control point* mean?
20. List several ways to minimize interference (cooperative use of frequencies).
21. What is the radiotelegraph distress call?
22. What is the radiotelephone distress call?
23. Under whose authority shall a distress call be sent?
24. What is the radiotelephone distress frequency in the medium frequency range?
25. What is the radiotelegraph distress frequency?
26. Who is eligible for a new license?
27. Who is not eligible for a license?
28. What is an ELT and how is it tested?
29. What is the purpose of an EPIRB transmitter?
30. When should the EPIRB batteries be replaced?
31. What are the following emission types? J3E, H3E, R3E, A3E.
32. Transmitters using F3E shall be capable of what amount of frequency deviation at 100% modulation?
33. When are transmitter frequency measurements required?
34. What is the frequency tolerance for ship transmitters?
35. What is the required frequency stability of aviation stations?
36. What operating authority does the General Radiotelephone Operator License convey?
37. How should a radio operator respond to inspections by foreign governments?
38. What is the term of a ship station license?
39. Explain the listening watch on 2182 kHz.
40. How long should station logs be retained?
41. List several things the log must contain.
42. What do you do if your license is lost?
43. What is a Marine Radio Operator permit?

44. When double sideband emmission is used, between what upper and lower percentage shall the modulation peaks be maintained?

45. What shall the nameplate contain? What type of transmitters have a nameplate?

46. Name three instances where it is not legal to operate a ship transmitter.

47. What is the procedure for initiating a call with another ship?

48. What are three procedure words and what do they mean?

49. Name several operator restrictions.

50. When should the phonetic alphabet be used?

51. If you are operating a portable ship unit, who may you communicate with?

52. What is the maximum power allowed in portable ship units (associated ship units)?

53. What are the station identification requirements of portable ship units?

54. Can associated ship units transmit from land?

55. Where should your operator license be posted?

56. In terms of priority of communications, what are the highest three?

57. What does the word *pan* mean?

58. What does the word *security* mean?

59. With whom may you share information you hear on the radio station receiver, when the transmission was not directed to you?

60. What is the silent period and when is it scheduled?

61. What types of communications may be done during the silent period?

62. What are the station identification requirements of ship stations?

63. Under what conditions must radio transmissions be suspended?

64. What is the proper procedure for testing station transmitters?

65. How often must tower lights be checked?

66. What class of license is required to make transmitter tests?

67. What transmitter range, in miles, is required on 2182 kHz under normal conditions?

68. What is *type acceptance* and how does it differ from "type approval"?

69. If you receive a notice of violation, you must make a written response within how many days?

70. Summarize the fines that may be imposed for violations of rules and regulations.

71. What is the amount of carrier suppression required in J3E transmissions?

72. Who sets the frequency standards?

73. How should an operator post his license if he is employed at more than one workstation?

74. Channel 16 is used for distress and safety calling. What is the frequency of this channel?

75. What channel is used for contacting the Coast Guard?

76. What is meant by the term duplex?

77. On what frequencies do ELTs operate?

78. Which classes of EPIRBs are mandatory on ships that travel more than 20 mi offshore?

79. Which class of EPIRB is normally used on ships that travel less than 20 mi offshore?

80. What is the effective radiated power requirement for class A, class B and class S EPIRBs?

81. What is the frequency tolerance requirement for Civil Air Patrol stations?

82. Which two licenses may not be held at the same time?

83. What is a working frequency?

84. If a person buys an aircraft, can he operate under the previous license, until he acquires one of his own?

85. Can stations in the aviation service be operated for brief periods of time without a station authorization (license) from the commission?

86. Can stations in the International Fixed Public Radiocommunications Services operate for brief periods of time with a station authorization from the Commission?

# 2
# Glossary

**absorption wavemeter**   A measuring device that consists of an *LC-tuned circuit*. A calibrated capacitor (C) is adjusted, and the coil (L) is loosely coupled to an oscillator or other source of radio frequency energy. When the capacitor is tuned to resonance, the meter or lamp indicates maximum. The wavemeter can be used to do any of the following:
1. Check tank circuit frequency
2. Check the field strength of an antenna
3. Check the output frequency of a transmitter
4. Locate the source of parasitic oscillations

**ammeter**   A device used for measuring electrical current. The range of the device may be increased by the addition of a parallel resistor (a shunt). An ac ammeter measures the effective current.

**angle modulation**   A modulator cannot change the frequency without changing the phase angle, or vice versa. Therefore, both FM (*frequency modulation*) and PM (*phase modulation*) are referred to as "angle modulation."

**antenna current and voltage relationships**   In a one-half wave antenna, the current is maximum and the voltage is minimum in the center. On the ends, the current is minimum and the voltage is maximum. (Please refer to chapters on antennas and transmission lines for details.)

**antenna length**   Primarily determined by its physical length. Resonant frequency may also be increased by adding a capacitance in series with the antenna. The resonant frequency can be decreased by placing an inductor in series with the antenna.

**atmospheric interference**  Below 30 MHz, radio signals reflect off the iono-sphere. This is what allows for long-distance communications. If a radio opera-tor is conducting local communications on a frequency below 30 MHz, long-distance stations might cause interference. Local communications are usually in the VHF or UHF ranges. Atmospheric static also tends to decrease as the frequency approaches 30 MHz.

**audio generator**  Can be used to align a receiver. Its signal appears at the detector stage of the receiver.

**automatic volume control**  Used to decrease the noise-to-signal ratio. It main-tains the volume at a preselected level. A voltage developed at the second detector is fed back to the grids of RF (radio frequency) and IF (intermediate frequency) amplifiers to adjust their gain. By adjusting the gain, the receiver volume is maintained at a constant level.

**azimuth**  The horizontal position extending 360° around the horizon.

**bandwidth**  The range of frequencies that are between the 3 dB (decibel) points of the resonance curve. As the Q (quality factor) of the circuit is increased, the bandwidth decreases. A low Q circuit has a wide bandwidth.

**bandwidth of emission**  All frequencies transmitted with power levels above a certain percent of the total radiated power (above 0.25%). It consists of the car-rier, sidebands, and harmonics.

**battery charge measurement**  The specific gravity of the sulfuric acid electrolyte is measured with a hydrometer. Specific gravity is the ratio of the weight of the electrolyte to the weight of water, as shown below. The specific gravity of dis-tilled water is 1.000. The specific gravity of a fully charged battery is about 1.300, which means that the electrolyte solution weighs 1.300 times more than an equal volume of distilled water. As the battery discharges, the sulfuric acid combines with the plates to form lead sulfate. As this happens, there is a greater percentage of water in the solution, therefore a lower specific gravity.

$$\text{Specific gravity} = \frac{\text{weight of electrolyte}}{\text{weight of water}}$$

**battery storage**  Lead acid batteries can be safely stored for long periods by drain-ing the electrolyte and refilling the battery with distilled water.

**buffer amplifier**  A low-gain amplifier that immediately follows the oscillator circuit in the transmitter. It shields the oscillator from load variations by providing it with a stable load. Without a buffer amplifier, changes in later RF amplifier or antenna circuits can cause changes in the frequency of the oscillator.

**capacitors in parallel**  The value of total capacitance of parallel capacitors is cal-culated in the same manner as total resistance of resistors in series. The capac-itance values simply and together as in: $C_1 + C_2 + C_3 =$ total capacitance.

**capacitors in series**   The value of total capacitance in series is calculated in the same manner as total resistance of resistors in parallel. For example, two capacitors of equal value when connected in series will have an equivalent value of one-half the original value of each capacitor. The equivalent of capacitors in series will always be less than the value of the lowest value capacitor in the circuit.

**carbon resistance microphone**   Operates on the principle of varying resistance. Sound vibrations induce movements in a diaphragm. Behind the diaphragm are carbon granules. When the diaphragm vibrates, the pressure varies on the carbon granules. The intensity of the current that flows through the carbon varies in accordance with the frequency and intensity of the sound waves that strike the diaphragm.

**carrier**   The output of a transmitter with no modulation present.

**cathode bypass capacitor**   A capacitor placed in parallel with the cathode resistor in a tube circuit. In a transistor circuit it is called the *emitter bypass capacitor* as it is placed in parallel with the emitter resistor. It prevents degeneration. If shorted, the resistor would be shorted and bias would change. If open, lower gain and lower output of the amplifier would be the result.

**cavity resonator**   A device that acts as a resonant circuit at microwave frequencies. Its resonant frequency is determined by its dimensions. It acts as an LC (inductive-capacitive) circuit at resonance.

**charge on a capacitor**   Lasts indefinitely unless purposely bled off or discharged.

**choke joint**   A joint must be used whenever two sections of waveguide are connected. A choke joint is often used instead of a simple-flange joint because it offers less signal loss. The choke joint has a circular one-quarter wavelength deep slot around it. Because of the slot, the RF energy entering the choke joint senses a short circuit across the mechanical connection. This allows the RF energy to move past the mechanical connection, as though it were not there, with minimal signal loss.

**circulator**   A ferrite device surrounded by a permanent magnetic field. It can be used as a duplexer to couple a receiver and a transmitter or two transmitters to a single antenna. The circulator provides a high degree of isolation between the transmitters and receivers.

**condenser (capacitor) microphone**   Operates on the principle of varying capacitance. Sound vibrations induce movements in a tightly stretched diaphragm that acts as one plate of a capacitor. The back plate of the microphone acts as the second plate of the capacitor. A fixed polarity voltage source is connected across these two plates. Changes in the position of the diaphragm cause a varying capacitance and a flow of electrons from one plate to the other. The voltage varies with the frequency and amplitude of the sound waves striking the diaphragm. Condenser microphones usually require a *pre-amp* (pre-amplifier) to make up for line losses.

**coupling between stages**   In a receiver, the degree of coupling affects the bandwidth, Q, and selectivity of the receiver. *Loose coupling* equals high Q, narrow bandwidth, with a high degree of selectivity in the receiver. Harmonic attenuation is greatest when loose coupling is used between coupled circuits. *Tight coupling* between stages equals low Q, wide bandwidth, with a low degree of selectivity in the receiver. Increasing the coupling between stages increases the bandwidth. If coupling is increased beyond a critical point, split-tuning will result.

**crystal**   Usually quartz is used. When an alternating current is applied across a crystal, it vibrates mechanically (*piezoelectric effect*). The crystal acts like, and therefore replaces, the tank circuit in an oscillator. The thickness of the crystal is the primary factor that determines its oscillating frequency. Temperature variations affect its frequency to a lesser degree. The oscillating frequency of a crystal can be slightly decreased by adding a variable capacitor in parallel with the crystal. (Remember that the crystal with its holder is equivalent to an LC tank circuit.) When two capacitors are connected in parallel, the total capacitance is increased. In a tank circuit, as the capacitance is increased, the resonant frequency decreases.

**directional wattmeter**   A wattmeter is placed in the transmission line to measure power. It has a switch that enables it to measure either *forward power* traveling from the transmitter to the antenna or *reflected power* traveling from an improperly matched antenna back to the transmitter. For example, if a directional wattmeter indicates 80 W forward power and 10 W reflected power, the actual transmitter output power is 70 W.

**doppler radar**   A radar signal is sent to a target. If the target is stationary, the same frequency will return. If the target is approaching, the frequency of the returning signal will be slightly higher. If the target is departing, the returning signal will be slightly less in frequency. This is because of the Doppler effect. (See Doppler Effect in Chapter 20).

**dummy antenna (dummy load)**   An artificial antenna that absorbs the RF signal from the transmitter. It allows the radio operator to test the transmitter without radiating any signals onto the airwaves. The dummy load consists of a non-inductive resistor in series with a capacitor. The resistor must have a resistance equal to the impedance of the antenna and a power rating equal to, or greater than, the output power of the transmitter.

**dynamic microphone**   Operates on the principle of a varying magnetic field. A lightweight coil mounted on a movable diaphragm is positioned within a permanent magnetic field. When sound vibrations induce movements in the diaphragm, voltage is induced in the moving coil as it cuts the field of the magnet.

This induced voltage varies in accordance with the frequency and amplitude of the sound waves that strike the diaphragm.

**dynamotor**   Used to step up dc voltage from low dc to high dc in portable installations. Output can be controlled by adjusting the battery input voltage with a rheostat.

**eddy currents**   Iron core losses in a transformer.

**electrolytic capacitor**   A capacitor that is capable of large values of capacitance. Proper polarity must be observed when connecting this type of capacitor in a circuit, or it can be destroyed.

**electromotive force (EMF)**   Also called difference of potential, voltage and *IR* drop. It refers to a force tending to place electrons in motion versus a static force.

**electrostatic field**   Refers to a static electric field, as between the plates of a charged capacitor. This field remains indefinitely until it is discharged.

**facsimile**   A method of communications where a fixed image is transmitted. At the receiver, the signal is converted to a permanent record. (The emission type had been designated A3C or F3C, formerly A4 and F4.) Maps, diagrams, text, etc., can be transmitted. (Also called *fax*.)

**ferrite beads**   Ferrites that can be threaded over wires in RF circuits. They act like radio-frequency chokes in that they exhibit a very low impedance to dc and a high impedance to radio frequencies. They are used for shielding and for suppression of parasitic oscillations.

**ferrites**   Specially constructed rods consisting of combinations of metallic oxides pressed together at high temperatures. When the rod is surrounded by a magnetic field, it exhibits the *ferromagnetic effect*. This allows a radio wave to travel through the ferrite in one direction but be strongly attenuated when traveling in the opposite direction. Ferrites are used in isolators and circulators.

**FET (field-effect transistor)**   A type of transistor that has very high input impedance.

**flywheel effect**   The quality of a tank circuit to generate a smooth sine-wave output in response to a distorted or pulsed input waveform. The flywheel effect makes possible the use of class B or class C operation in RF amplifiers. For example, the class B amplifier amplifies only one half of its input sine wave. When these half waves enter a resonant tank circuit, they emerge as restored sine waves. Thus, distortions and undesired harmonics are minimized.

**frequency doubler**   Its output is twice its input frequency. Doublers are usually operated as class C amplifiers because of the rich harmonic content. To form a frequency doubler, you must reduce the value of either the capacitor or inductor in the tank circuit to one fourth of its original value.

**frequency multiplier**  Usually operated as class C because of the high distortion and harmonic content. This is one of the rare cases where high distortion and harmonics are desirable. The output tank circuit can be tuned to resonate to any one of the harmonics present. If tuned to the second harmonic, a doubler is formed. If tuned to the third harmonic, a tripler is formed.

**glideslope**  Part of the ILS (instrument landing system). The glideslope provides vertical positioning information during the approach. (See Localizer.)

**grid-dip meter**  A measuring device that can be used to measure the frequency of a non-operating tank circuit. It consists of an oscillator with a coil that can be loosely coupled to the tank circuit being measured. A variable capacitor is adjusted until the grid-dip oscillator is tuned to the resonant frequency of the tank circuit. At this time, the tank circuit absorbs energy from the oscillator, and the grid-dip meter dips to a minimum reading.

**grid-leak bias**  A common type of bias used in vacuum tubes. The current flows through the grid-leak resistor towards ground. If an oscillator is oscillating, there will be grid-leak bias present.

**grounded grid amplifier**  A vacuum tube RF power amplifier that generally does not require neutralization.

**harmonic**  An even multiple of a fundamental frequency. For example, the seventh harmonic of 100 kHz is 700 kHz. Push-pull amplifiers tend to cancel out even harmonics. Square waves consist of numerous odd harmonics. Triangular waves consist of numerous even harmonics.

**hertz**  The hertz (Hz) is the basic unit for measurement of frequency. One hertz is the same as 1 cps (cycle per second). Another way of expressing one complete cycle is $2\pi$ rad.

**image frequency**  An undesired signal generated by superheterodyne receivers. It is equal to the LO (local oscillator) frequency plus the IF (intermediate frequency). It is also equal to the incoming signal frequency plus two times the IF. For example, an FM receiver has a 10.7 MHz IF. If the receiver is tuned to 100 MHz, the image frequency would be 121.4 MHz (100 MHz plus 10.7 MHz plus 10.7 MHz). If a powerful transmitter in close proximity to the receiver transmits on the image frequency, that signal can be heard on the FM radio. This is because when 121.4 MHz is mixed with 110.7 MHz (LO frequency), a difference frequency of 10.7 MHz is produced. This easily travels down the IF amplifier stages.

**impedance**  The alternating-current version of resistance. Impedance is also measured in ohms.

**impedance matching**  For maximum transfer of power, the impedance must be matched between circuits or stages. For example, the impedance of the transmitter must match the impedance of the transmission line, which must match the impedance of the antenna. If a mismatch exists at any point, power losses will result.

**inductance of a coil**   Inductance varies directly to the square of the number of turns on the coil. If an air core is replaced with an iron core, the inductance will increase.

**infinite transmission line**   A theoretical transmission line that is infinitely long. Such a line has a characteristic (*surge*) impedance that is the same at any point in the line. If a finite transmission line is terminated in a resistance equal to its surge impedance, the line is characterized by its input impedance equaling its output impedance. When a line is terminated in this manner, it looks like an infinitely long line. Such a line has no standing waves.

**integrated circuit (IC) pin numbering**   Looking at the top of an IC, you can see a dot or a notch on the case. Start counting the pins from that point counter-clockwise around the IC. Of course, if you are looking at the bottom of the IC, you would number the pins clockwise.

**isolator**   A ferrite device that is surrounded by a magnetic field. When radio frequency energy enters the device from one direction, no attenuation takes place. When energy enters the device from the opposite direction, it is absorbed. It is a one-way valve for radio waves. Uses include the attenuation of reflected waves in transmission lines while passing the forward power from the transmitter to the antenna. It can be used as a buffer between a microwave oscillator and waveguide by protecting the oscillator from load variations. It can also be used to help prevent intermodulation by preventing external signals from entering transmitters through their antennas. This is especially important in locations where several transmitters are in close proximity.

**kilowatt-hour**   The unit of electrical energy equal to 1 kW of energy expended for 1 h. It is measured with a watt-hour meter.

**knife-edge refraction**   The reduction of atmospheric attenuation of a radio wave when it passes over a sharp object like a building or mountain ridge. A refraction, or bending, of the radio wave occurs.

**lead length**   Leads should be kept as short as possible when working with RF circuits. This minimizes stray capacitance. The higher the frequency, the more critical it becomes.

**limiter**   A low-gain and constant-output amplifier used in FM receivers. It is placed immediately before the discriminator (detector) to remove static and amplitude variations from the FM signal.

**link coupling**   Used to couple widely separated circuits. It provides some harmonic attenuation because it reduces capacitive coupling between the stages. One application is the coupling of the RF tank circuit to the antenna-matching network.

**local oscillator (LO)** A radio-frequency oscillator that oscillates slightly above the frequency of the incoming signal in a receiver. Its operating frequency is determined by the IF of the receiver. When the IF is 455 kHz, as in AM receivers, the LO operates at 455 kHz above the incoming signal. When the IF is 10.7 MHz, as in FM receivers, the LO operates at 10.7 MHz above the frequency of the incoming signal.

**localizer** That part of the ILS that provides information about aircraft horizontal positioning during its approach.

**loran C** A long-distance radio-navigation system. Because it operates in the low frequency band, on 100 kHz, it uses ground-wave propagation. The operation is based on the simultaneous transmission of synchronized pulses by widely separated transmitting stations. The loran receiver on the ship or aircraft analyzes signal time delays to determine position.

**magnetron tube** A device used at microwave frequencies; it is a type of diode and must be surrounded by a strong magnetic field. It generates SHF (super high frequency) signals.

**marconi antenna** A vertically polarized antenna. One quarter wavelength acts as a radiator. The second half is grounded. There are two methods of feeding such an antenna:

    *series-fed marconi* The antenna is insulated from the ground and elevated to the desired height above the ground. The second half is in the form of several radials—quarter-wave conductors that extend perpendicularly from the base of the antenna.

    *shunt-fed marconi* The antenna is not insulated from the ground. Therefore, the dc resistance to ground is zero. Impedance matching is accomplished by tapping the antenna at the proper point above ground. This is possible because the impedance is very low at ground level but steadily increases at points nearer the top.

**mixer (First detector)** Produces sums and differences of its input signals. Its two inputs are the incoming RF signal and the local oscillator signal. One of its output signals becomes the IF of the receiver.

**modulation index (mi)** The ratio between the modulating frequency and the frequency deviation. It is directly proportional to frequency and is expressed as follows:

$$\text{Modulation index} = \frac{\text{radio frequency } (RF)}{\text{audio frequency } (AF)}$$

where
    $RF$ = carrier frequency deviation
    $AF$ = audio modulating frequency

**modulation systems** Two basic systems of modulation are AM and FM. Although FM occupies more bandwidth, the reception is clearer because static interference is composed of amplitude variations. Static, therefore, has a greater effect on AM radio reception.

**nautical mile (n mi)**   This is the fundamental unit of distance used in navigation. It is equal to 6080 ft. One n mi equals 1.1516 mi. One knot equals 1 n/mi/h. (See statute mile.)

**neutralization**   A grid-dip meter or neon bulb can be used to check for the necessity of neutralization. When the excitation voltage is removed from an RF amplifier and RF is still present, the amplifier requires neutralization. To begin, the B+ (plate voltage) is removed, with all other voltages applied. A neutralization capacitor is then adjusted until the RF is absent. Neutralization is accomplished when the net capacitive feedback voltage from plate to grid is zero and when no current flows in the plate circuit.

**nitrogen gas**   Used in some transmission lines and waveguides for the purpose of reducing moisture in the line.

**noise levels**   Because most static and other forms of noise are amplitude modulated, FM receivers have a principal advantage in that they eliminate all amplitude variations.

**noninductive resistor**   Because resistors at ultra-high frequencies might appear as a capacitance or inductance, special resistors must be used at these frequencies. Use noninductive resistors in series with grid or plate when trying to eliminate parasitics.

**ohmmeter**   An electrical measuring device that is used for determining electrical resistance. It is also commonly used to determine whether a circuit is open or shorted.

**omnidirectional antenna**   An antenna that receives or radiates equally well in all directions. It has no directivity.

**op-amp (operational amplifier)**   An amplifier whose characteristics are determined by components external to the amplifier. For example, the gain of an inverting op-amp is determined by the ratio of the feedback resistor (Rf) and the input resistor (Ri). When the input voltage is applied to the inverting input the minus sign in the formula indicates that there is a phase reversal at the output of the op-amp. Figure 2-1 shows an inverting op-amp and the method of calculating its gain.

**parasitic element**   An antenna element placed a fraction of a wavelength away from the active element. It increases the directivity and unidirectionality of the antenna. When several parasitic elements are used, a directional antenna is formed with most of the RF energy being directed from the front of the antenna. Minimal energy is sent or received from the back or the sides of the antenna. The antenna directs the radio waves in much the same way as a spotlight directs light.

$$\text{gain} = -\frac{R_f}{R_i} \quad \text{where:} \quad \begin{aligned} R_f &= \text{feedback resistance} \\ R_i &= \text{input resistance} \end{aligned}$$

**2-1**  Gain calculations in an inverting op-amp (operated amplifier).

**parasitics**  Spurious oscillations at undesired frequencies. Although they are most common in RF amplifiers, they can originate in any circuit. They can be produced by stray capacitance or inductance from excessively long leads. Proper shielding and short leads are necessary to prevent them. Series noninductive resistors or RF chokes can be placed in the plate and grid leads. They can be located with an absorption wavemeter or heterodyne frequency meter.

**peak inverse voltage (PIV)**  The maximum reverse voltage that can be applied to a diode before it is damaged. Semiconductor diodes are more sensitive to reverse voltage than are vacuum-tube diodes.

**percent modulation**  Modulation is the method by which the audio is superimposed onto the RF carrier. The larger the modulating voltage, the greater the percent of modulation. Modulation of 100% offers the best signal-to-noise ratio and intelligibility at the receiver. If 100% is exceeded, distortion will result. Percent modulation should be between 70% and 100%.

**phase relationship in capacitive circuit**  Remember "ELI the ICE man." In a purely capacitive circuit, the current ($I$) leads the voltage ($E$) by 90°. As resistance is introduced into the circuit, the phase angle decreases. In a purely resistive circuit, there is no phase delay; current and voltage are in phase.

**phase relationship in inductive circuit**  Remember "ELI the ICE man." In a purely inductive circuit, the voltage ($E$) leads the current ($I$) by 90°. As resistance is introduced into the circuit, the phase angle decreases to zero when pure resistance is present.

**plate distortion**  Usually is high in class B or class C RF amplifiers. This distortion presents no problem because the flywheel effect of the tank circuit reproduces a pure sine wave. This characteristic is used in frequency multiplier stages.

**power consumption**  In an ac circuit, power consumption depends on the effective value of $E$ and $I$. $P = E \times I$.

**pre-emphasis**  A method of increasing the signal-to-noise ratio by selectively attenuating low frequencies and passing the higher audio frequencies. This is accomplished because a capacitor offers less reactance to the higher frequencies. The lower frequencies are forced to be attenuated as they pass through an RC network. This is restored to normal in the FM receiver when the signal passes through a de-emphasis circuit.

**push-pull operation**   Amplifier operation when even harmonics are canceled out. It is especially helpful in class B audio amplifiers, because both halves of the input waveform are amplified, which reduces the distortion.

**Q (quality factor)**   Gain or efficiency of a circuit. It varies with the amount of coupling and the amount of resistance. As the coupling is loosened, the Q increases. As the resistance is decreased, the Q increases. As the Q is increased, the bandwidth decreases.

**radar mile**   Radio waves travel 1 n mi in 6.2 $\mu$s. A radar mile is 12.4 $\mu$s. This is the time required for the radar pulse to travel 1 n mi to a target and 1 n mi back to the radar receiver.

**reactance tube modulator (reactance tube)**   Often used in direct FM modulation. Its output appears as a varying capacitance or inductance to an oscillator.

**reflectometer (SWR meter)**   A device placed in the transmission line to measure the standing wave ratio. A bridge circuit is used to compare the forward and reflected power in the line.

**regulation (of power supply or generator)**   The relationship of the no-load and full-load voltage output. The output voltage of a well-regulated power supply will not vary with changes in load conditions.

**residual magnetism**   The polarization of a magnetic material with the magnetizing force removed. It is the molecular alignment of a permeable material.

**resistance versus gauge of wire**   The lowest gauge number is the largest wire and therefore has the lowest resistance. Comparing 12-gauge, 10-gauge, and 4-gauge wire, the 4-gauge wire has the lowest resistance because it is the largest.

**resistance versus wire size**   The resistance varies inversely with the cross-sectional area of the wire conductor. Therefore, if the area is doubled, the resistance is halved. If the area is tripled, the resistance is one third of the original value (2 times the area equal $1/2$ the resistance; 3 times equals $1/3$; 4 times equals $1/4$, etc.).

**resonance**   The condition in a series or parallel ac circuit where the inductive and capacitive reactances are equal. The two reactances cancel each other out, leaving only pure resistance in the circuit.

**RFC (radio frequency choke)**   Passes dc and low frequencies but attenuates high frequencies. It can be used between a motor-generator and RF transmitter to prevent transmitter RF from being fed back. This prevents insulation breakdown. They are also used in power supply leads to prevent the entry of RF into the power supply. They provide dc coupling and ac isolation.

**ribbon microphone**    A microphone with a loosely suspended conductor in a strong magnetic field.

**root-mean-square (rms)**    A method of expressing the amplitude of a sine wave. It is also called the *effective value*. A light bulb will be equally bright when connected to a 100-V dc source or a 100-V rms ac source. Most ac meters measure rms values.

**selectivity of a receiver**    This is the ability of the receiver to select the desired signals and attenuate adjacent signals. Selectivity is greatest when loose coupling is used between stages.

**self-excited generator**    A generator in which the armature supplies current to the field coils. The generation of its output voltage depends upon the residual magnetism in the field magnets. These *shunt-wound generators* have a good voltage regulation, because they produce fairly constant output during load variations.

**series dc motor**    Speed is affected primarily by its load.

**sideband power at 100% modulation**    Power in the upper sideband is one sixth the total power. Power in the lower sideband is also one sixth total power. Therefore, the total sideband power is one third the total power when both sidebands are considered.

**signal generator**    A precision test instrument that produces a sine wave of variable frequency and amplitude. It is used for calibrating other test equipment and aligning receivers. For example, it can be connected to the control grid of the RF amplifier to test its function.

**signal-to-noise ratio**    The ratio of signal strength to background noise strength. A good signal-to-noise ratio will result in a clear signal at the receiver. It can be maximized by using:
- High percentage of modulation in the transmitter (not to exceed 100%)
- Pre-emphasis in the transmitter
- De-emphasis in the receiver
- Narrow band pass in the receiver

**single-sideband suppressed carrier (SSSC)**    A type of emission where either the upper or lower sideband is filtered out and the carrier is suppressed or balanced out. The amount of suppression required is 40 dB.

**skin effect**    The tendency for radio frequency energy to travel on the surface of a conductor, rather than through the entire cross-sectional area. The skin effect increases as the frequency increases.

**soldering tips**   *Soldering* is the joining of two metals together by causing a low melting point metal alloy (solder) to flow over the metals. When the solder solidifies, a bond is formed.

Solder consists of small hollow wire made of a metal alloy. The alloy is a combination of lead and tin. The hollow center is filled with a nonacidic *flux* for soldering electronics circuits.

Because the flux has a lower melting point, it flows over the connection before the solder flows. The flux cleans and prepares the surface by removing any oxidation, oil, grease, or organic films. This promotes more efficient wetting and a better solder connection.

*Wetting* can be a measure of how well the materials can be soldered. *Wetting* refers to how easily the molten solder flows and wets the surface of the metals being soldered. Good wetting means the molten solder flows easily, in a free-flowing fluid manner, over the metals being joined—even into the small cracks and crevices. Poor wetting can be caused by a dirty surface. When poor wetting occurs, the solder does not adhere securely to both metals, resulting in a poor bond and a poor electrical connection. Insufficient heat from the soldering iron can also prevent proper wetting.

The following are some things to keep in mind while soldering:
1. Thoroughly clean both surfaces
2. Make the connections as tight as possible
3. Heat the connections before applying the solder
4. Make sure you apply enough heat to ensure a free-flowing wetting of the surfaces.
5. After the wetting has taken place, remove the heat and hold the wires stationary until the solder cools. Cooling is indicated by a dulling of the shine on the surface of the hot solder. If the joint is moved too soon, a cold solder joint can be formed. Colder solder joints make weak and inefficient electrical connections.
6. Use a heat sink to dissipate some of the heat so that a sensitive component will not be changed. Place the heat sink between the joint being soldered and the component. You can use needle-nose pliers as a heat sink.

**split tuning**   As coupling between stages is increased up to a critical point, the bandwidth becomes wider. When the critical point is exceeded, two resonant peaks form. This condition can be eliminated by reducing (loosening) the coupling.

**square wave**   A fundamental frequency plus numerous superimposed odd harmonics.

**squelch**   A receiver circuit designed to quiet the receiver when no signal is present. It works in conjunction with the AVC (automatic volume control) circuit. The advantage of a squelch is that the receiver remains quiet when the frequency is clear. The disadvantage is that weak signals that are near the background noise level might not trigger the circuit. When listening for a very weak signal, it is often necessary to turn the squelch off.

**stacked antennas** Stacking of antennas increases the directivity of the antenna system. When the antenna is used for receiving, the reception is increased. When it is used as a transmitting antenna, the effective radiated power is increased.

**standing waves** Variations in the voltage and current along a transmission line. They are caused by an impedance mismatch between the line and the antenna. When the surge impedance of the transmission line is equal to the impedance of the antenna, there are no standing waves on the line, and maximum power is transferred. When there are no standing waves, the rms voltage along the RF transmission line is the product of the surge impedance and the line current ($E = I \times Z$). The SWR (standing wave ratio) is maximum current divided by minimum current or maximum voltage divided by minimum voltage at any point along the line. The optimum SWR is a 1-to-1 ratio. At this ratio, the current and voltage are the same at any point along the line and there are no standing waves.

**statute mile (mi)** This is equal to 5280 feet or 0.8684 n mi.

**surge impedance** The surge impedance, also called characteristic impedance, of a transmission line is determined by the size and spacing of the conductors and the dielectric material between them. A common surge impedance for coaxial cable is 50 Ω. A common value for parallel transmission line is 300 Ω. The surge impedance of the line must be matched to the impedance of the antenna for maximum power transfer.

**tank circuit** A tank circuit is a parallel resonant circuit placed at the output of an RF amplifier or frequency multiplier. It consists of a capacitance and inductance in parallel. The resonant frequency is determined by the values of the capacitor and coil.

**time constant** When a resistor, capacitor, and battery supply are connected in series, one time constant is the time required for the capacitor to charge through the resistance to 63.2% of the battery voltage. In five time constants, the capacitor will charge to the full supply voltage. If the three components are connected in parallel, the capacitor will charge immediately to the supply voltage. But when the battery is removed, the capacitor and resistor are connected in series and the capacitor discharges through the resistor. In one time constant a capacitor will discharge 63.2% which is 36.8% of the supply voltage. In five time constants, the capacitor will fully discharge. Time constant is equal to resistance times capacitance.

**transistor bias** The emitter-base junction is forward biased and the collector-base is reverse biased. An easy way to remember battery polarities is:

A battery has a p (positive) and an n (negative)

A transistor has p (p-type material) or n (n-type material)

*forward bias* p is connected to p, and n is connected to n.

*reverse bias* p is connected to n, and n is connected to p.

Forward bias of emitter-base in pnp and npn transistors:

$$+ \; p \; n \; p \qquad - \; n \; p \; n$$
$$- \qquad\qquad +$$

Reverse bias of collector-base in pnp and npn transistors:

$$p \; n \; p \; - \qquad n \; p \; n \; +$$
$$+ \qquad\qquad -$$

The emitter-base bias is considerably less than the collector-base bias. If they are equal, the circuit will not operate. An example of proper bias and improper bias voltages is:

proper                  improper

p n p                  p n p
2 V    12 V          12 V    12 V

**transistor circuit configurations**    Three basic circuit configurations are used. They are the common base, common emitter, and common collector circuits. It is important to know their input and output impedance characteristics (discussed in another chapter). It is also important to know that of the three circuits, only the common emitter produces a phase reversal. That is, the amplified output signal is 180° out of phase with the input signal.

**transistor damage**    Transistors are very sensitive to heat. The increased temperature causes the collector current to increase. This will eventually damage the transistor if it is allowed to continue.

**transmitter power calculations:**
*direct method*    The antenna resistance and current are measured at the same point. Use ohm's power law as: power equals antenna current squared times antenna resistance.
*indirect method*    The plate current and plate voltage of the final RF amplifier stage is determined. Power is determined by the product of the two figures times an efficiency factor. The transmitter power can be increased by increasing either plate current or plate voltage of the final amplifier.

**traveling wave tube (TWT)**    A UHF amplifier tube. The input signal to be amplified is inserted at the cathode end of the helix (end closest to the electron gun). The coupling in and out is accomplished with a waveguide transducer.

**trippler**    A frequency tripler has an output equal to three times the input frequency. This can be accomplished by tuning its tank circuit to the third harmonic of the amplified input signal.

**unidirectional**    Refers to a directional antenna system. Unidirectional means that the antenna transmits or receives with greater efficiency in one direction. This is contrasted with an omnidirectional antenna that receives or transmits equally well in all directions.

**UTC (Universal Time Coordinated)**    A standardized method of telling time. It was formerly called GMT (Greenwich Mean Time) because the zero meridian passes near Greenwich, England. To calculate UTC time, simply add 8 hours to PST (Pacific Standard Time). During Daylight Saving Time (usually from the last Sunday in April to the last Sunday in October) add only seven hours to the PDT (Pacific Daylight Time). Therefore, if it is 1300 hours in California in the month of August, it is 2000 hours UTC. If it is 1800 hours (6 P.M.) in Oregon in the month of December, the UTC time would be 0200 hours the next day.

**valence electrons** Outer electrons of the atom. They are responsible for electrical conduction.

**varactor** Also called a voltage-variable capacitor, Varicap, varactor diode and tuning capacitor. A capacitance exists at a pn junction. A varactor diode is constructed with special impurities to enhance that capacitance. As the reverse bias on the diode increases, the depletion region of the junction (acting like the dielectric of a capacitor), widens. As it widens, the two plates (the p and n sections) move apart and the capacitance decreases. Varactors are used in frequency multipliers, electronic tuning, and in FM circuits.

**varistor** Any device with an electrical resistance that can be changed by variations in either voltage, current, or polarity. Semiconductor diodes and transistors can be considered as varistors.

**vertical antenna** An antenna that is placed perpendicular to the ground. It has an omnidirectional radiation pattern, radiating or receiving equally well in all horizontal directions.

**very low frequency (VLF)** Includes 3 to 30 kHz signals. Transmission on these frequencies makes communication with submarines possible.

**voltage doubler** A circuit in which two capacitors are charged in parallel and discharged in series. The arrangement results in twice the voltage at the output.

**VTVM (vacuum-tube voltmeter)** A sensitive voltmeter that has a high input impedance. Because of this, it can be used to measure small voltages without loading the circuit.

**watt-hour meter** Used to measure electrical energy.

**wattmeter** Device used for measuring electrical power. It calculates voltage times current and corrects for phase differences. The wattmeter measures true power.

$$P = I \text{ (current)} \times E \text{ (voltage)}$$

**waveguide** A hollow metal tube used for the transmission of microwaves. It is not used at lower frequencies because its size would be prohibitive.

**wavelength** A wavelength is the distance a radio wave will travel during the time required for one cycle. It is inversely proportional to frequency.

**zener diode** A diode used for voltage regulation. When an unregulated voltage is applied to the diode, a fixed reference voltage is produced. As the reverse bias is increased, the diode will reach a point where its peak reverse voltage is exceeded. At this point the diode breaks down and current flows. Zener diodes have different breakdown voltages, depending on their design.

**zero beat**   When two signals are mixed, or heterodyned, the sum and difference frequencies are produced. As the two frequencies approach the same frequency, their difference frequency (beat frequency) becomes lower. When the two original frequencies are equal, there will be no audio beat note produced. This is called a zero beat. Zero beats are used in the heterodyne frequency meter.

# Study questions
# Glossary of electronics

1.  What three uses does an absorption wave meter have?
    a. _____
    b. _____
    c. _____

2.  What precaution should be taken when using the wave meter? _____

3.  An ac ammeter measures what kind of current? _____

4.  What are the voltage and current relationships on a one-half wave dipole antenna,
    a.  at the center?_____
    b.  at the ends?_____

5.  The resonant frequency of an antenna can be lowered by making the antenna _____ or by adding a series _____.

6.  At what point in a receiver is the signal from an audio-signal generator heard? _____.

7.  In the AVC circuit, a signal from the second detector is fed back to the _____ to control the gain of the receiver.

8.  As the Q of a circuit is increased, the bandwidth is _____.

9.  Bandwidth of emission is defined as all transmitter frequencies with power levels above a certain percent of the _____.

10.  A _____ can be used to determine the charge of a battery.

11.  If a battery must be stored for a long time, the electrolyte should be drained and replaced with _____.

12. A buffer amplifier provides the oscillator with a _____ and a _____ load.

13. Capacitors in series act like resistors in _____.

14. Capacitors in parallel act like resistors in _____.

15. The carbon resistance microphone operates on the principle of a varying _____.

16. If a cathode bypass capacitor is damaged, the output of the amplifier would be _____ and _____.

17. How long does a charge remain on a capacitor? _____.

18. RF chokes can be used in series with the leads of a motor generator in order to protect the _____ from _____ damage.

19. Knife-edge refraction is the reduction of atmospheric attenuation of a radio wave when it passes over a _____ object like a building or mountain ridge. A _____ of the radio wave occurs.

20. Of the three basic transistor configurations, which is the only one that produces a phase reversal at its output? The common _____ configuration.

21. The common base configuration offers a _____ input impedance and _____ output impedance.

22. The commutator acts like a mechanical _____.

23. The condenser microphone requires a separate power supply and a _____ to make up for line losses.

24. To obtain a high degree of selectivity in a receiver, the coupling between stages should be _____.

25. The frequency of a quartz crystal can be lowered by adding a capacitor in _____.

26. The frequency of a quartz crystal can be increased by adding a capacitor in _____.

27. Crystal oscillators are very sensitive to _____ variations.

28. Downward modulation is caused by _____ grid bias on the final stage.

29. A dummy antenna, used for testing the transmitter, consists of a _____ and _____ in series.

30. Dynamic instability can be caused by poor _____.

31. The dynamic microphone consists of a movable coil in a _____ magnetic field.

32. The dynamotor is used to _____ dc voltage.

33. How is dynamotor output controlled?_____.

34. Eddy currents are _____ losses in a transformer.

35. What class amplifier has the highest efficiency? Class _____.

36. What class amplifier has the lowest efficiency? Class _____.

37. What class amplifier has the highest fidelity? Class _____.

38. Facsimile is a method of communications where a _____ image is transmitted. This is converted into a permanent record at the receiver.

39. The FET (field-effect transistor) transistor is characterized by a very _____ input impedance.

40. Filament polarity should be reversed periodically because this _____.

41. The flywheel effect removes the _____ present in a class B or class C amplifier.

42. Frequency doublers are usually operated as class _____.

43. Because of high harmonic content, plate distortion is necessary for operation of the _____ circuit.

44. A grid-dip meter is used to measure the _____ of a tank circuit.

45. In a grid-leak bias, the current flows through the grid-leak resistor toward _____.

46. Grid-leak bias present indicates that an oscillator _____.

47. A grounded grid amplifier usually does not require _____.

48. The seventh harmonic of 100 Hz is _____.

49. When the frequency of a heterodyne frequency meter equals the frequency of the transistor, a _____ results.

50. Proper impedance matching results in transfer of _____ to the load.

51. The inductance of a coil varies directly with the square of the _____ of _____ on the coil.

52. An infinite transmission line is characterized by _____ being equal to _____.

53. In a communications receiver, the mixing of the incoming signal and the local oscillator signal results in the production of the _____ frequency.

54. A radio tuned to a local FM station might receive interference from a nearby transmitter. This is called _____.

55. The kilowatt-hour is the unit of _____ _____, and is measured by the _____ meter.

56. When working with RF circuits, lead length should be _____.

57. In an FM receiver, the limiter is a _____ gain and _____ output device that removes the static from the FM signal.

58. Link coupling is used to couple _____.

59. In a shunt-fed Marconi antenna, the dc resistance to ground is _____.

60. The ratio between the modulating frequency and the frequency deviation is called _____.

61. A grid-dip meter or neon bulb can be used to check for the necessity of _____.

62. To neutralize an amplifier, the _____ is removed with all other voltages applied.

63. Nitrogen gas is used in some transmission lines to prevent _____.

64. The carbon resistor can appear as an _____ and _____ at ultra high frequencies.

65. The _____ is a device that can be used to determine if a circuit is shorted or open.

66. An antenna that receives or radiates in all directions is an _____ antenna.

67. A parasitic element in an antenna increases the _____ and _____ of the antenna.

68. Semiconductors are more sensitive to _____ than tubes.

69. The percent of modulation should be between _____ and _____ percent.

70. In a purely capacitive circuit, the _____ leads the _____ by 90°.

71. In a purely inductive circuit, the _____ leads the _____ by 90°.

72. A high degree of plate distortion presents no problem in RF amplifiers because of the _____ effect of the tank circuit.

73. Power consumption in an ac circuit depends on the _____ value of voltage and current.

74. Push-pull tends to eliminate _____ harmonics.

75. A reactance tube modulator in an FM modulation system appears as a varying _____ or _____ to an oscillator.

76. Regulation is the relationship between the _____ and the _____ voltage.

77. Residual magnetism is the polarization of a magnetic material with the magnetizing force _____.

78. As the gauge of wire becomes smaller, the resistance of the wire becomes _____.

79. If the cross-sectional area of a conductor is tripled, the resistance would be _____.

80. At resonance, _____ and _____ are equal.

81. An RFC can be used in a motor-generator circuit to prevent _____ breakdown.

82. A ribbon microphone is a microphone with a _____ suspended conductor in a _____ magnetic field.

83. Selectivity of a receiver is its ability to select wanted signals and _____ _____ _____.

84. Selectivity is greatest when there is _____ coupling between the stages.

85. The speed of a series dc motor is determined by the _____.

86. At 100% modulation, the power in the upper sideband is _____ of the total power.

87. SSSC transmission is a method of communication where one set of sidebands is _____ and the carrier is _____.

88. Skin effect is the tendency of RF to travel at the _____ of a conductor.

89. A square wave consists of a sine wave with many _____ harmonics superimposed.

90. Stacked antennas increase the _____ of an antenna system.

91. Standing waves result when there is an _____ mismatch.

92. Time constant in a RC circuit is the time required for the capacitor to charge up to _____ percent of the applied voltage.

93. In a transistor, the emitter-base is _____-biased and the collector-base is _____ biased.

94. The potential applied to the emitter-base is considerably _____ than that applied to the collector-base.

95. If a transistor is overheated, damage can result when the _____ current increases beyond acceptable limits.

96. The direct method of measuring transmitter power is measuring the antenna _____ and _____, then Ohm's law is applied.

97. In a TWT, the input signal is applied at the _____ end of the helix.

98. Valence electrons are responsible for _____ _____.

99. In a varactor, the _____ varies as the _____ across the diode varies.

100. A VTVM is a sensitive meter with a _____ input impedance. As a result of this quality, it does not _____ the circuit.

101. A wattmeter calculates _____ times _____ and corrects for _____ _____.

102. A wattmeter measures _____ power.

103. The watt-hour meter is used to measure _____.

104. Wavelength is inversely proportional to _____.

105. A waveguide is a hollow metal transmission line. It is not used at low frequencies because of _____.

106. Which two emission types are considered to be angle modulation?

107. What does the term *azimuth* refer to?

108. Why are choke joints used instead of flange joints?

109. The glideslope provides _____ positioning information during the aircraft approach.

110. Inserting an iron core into an air core will cause its inductance to _____.

111. How are integrated circuit (IC) pins numbered?

112. On what frequency does loran C operate?

113. A nautical mile is equal to _____ microseconds.

114. A radar mile is equal to _____ μs.

115. What does the term *wetting* mean, in relation to soldering?

116. What two conditions might prevent adequate wetting of a solder joint?

117. If it is 6 P.M. in Oregon in the month of December, what is the UTC time?

118. Very low frequencies are used for communication with submarines. What frequencies does this band include?

# 3
# Emission types and frequency ranges

All radio communication signals can be described in terms of their basic characteristics, how much space they occupy, and the frequency band on which they are transmitted. The *emission type* describes the characteristics of the radio signal. The *necessary bandwidth* describes how space the signal occupies. *Frequency range* describes where on the radio-wave spectrum the radio signal is transmitted.

The Federal Communications Commission has devised a unique system of describing the emission types and necessary bandwidth of radio signals. This information is detailed in the Code of Federal Regulations, Title 47, Part 2, Subpart C—Emissions.

## Emission types

The Federal Communications uses a standardized method to classify emission types. Three symbols are used to describe the basic characteristics of radio waves. Each symbol tells you something about the radio signal, as follows:

**First symbol**  Describes the type of modulation of the main carrier:
1. Emission of an unmodulated carrier                                      N
2. Emission with amplitude-modulated main carrier:
   Double sideband                                                        A
   Single sideband, full carrier                                          H
   Single sideband, reduced carrier                                       R
   Single sideband, suppressed carrier                                    J
   Independent sidebands                                                  B
   Vestigial sideband                                                     C
3. Emission in which the main carrier is angle modulated:

| Frequency modulation | F |
| Phase modulation | G |

*NOTE:* Whenever frequency modulation F is indicated, phase modulation G is also acceptable.

4. Emission in which main carrier is amplitude and angle modulated. — **D**

5. Emission of pulses:

| Sequence of unmodulated pulses | P |

A sequence of pulses:

| Modulated in amplitude | K |
| Modulated in width/duration | L |
| Modulated in position/phase | M |
| Where carrier is angle modulated during the period of the pulse. | Q |
| Combination of the above | V |

6. Cases not otherwise covered — **X**

**Second symbol**   Nature of signal(s) modulating the main carrier:

1. No modulating signal — **0**
2. A single channel with digital information without the use of a modulating subcarrier — **1**
3. A single channel with digital information with use of a modulating carrier — **2**
4. A single channel containing analog information — **3**
5. Two or more channels containing digital information — **7**
6. Two or more channels containing analog information — **8**
7. Composite system with one or more channels containing digital information plus one or more channels containing analog information — **9**
8. Cases not otherwise covered — **X**

**Third symbol**   Type of information to be transmitted:

1. No information transmitted — **N**
2. Telegraphy—for aural reception — **A**
3. Telegraphy—for automatic reception — **B**
4. Facsimile[1] — **C**
5. Data transmission, telemetry, telecommand — **D**
6. Telephony—also called voicE or phonE[2] — **E**
7. Television (video)[3] — **F**
8. Combination of the above — **W**
9. Cases not covered — **X**

Table 3-1 summarizes some of the most commonly used emissions.

---

[1]*HINT:* The facsimile receiver produces a permanent Copy of the image that is sent to it. Remember C for copy. Any designation ending with a C is facsimile, such as A3C and F3C.

[2]*HINT:* Both voicE and *phoneE end with the letter E. Please note that signals designations that pertain to voice communications also end with an E, such as: A3E, J3E, G3E, R3E, and H3E.

[3]*HINT:* To remember the third symbol for TV, remember F for focus.

## Table 3-1. Emission types.

| Amplitude modulation | Emission types | Frequency modulation |
|---|:---:|---|
| NON ———————— | *Unmodulated carrier* | ———————— NON |
| A1A ———— On-off keying | *Telegraphy* | Frequency shift keying —— F1B |
| A3E ———— (Double sideband, full carrier) | *Telephony* | Phase modulation ———— G3E |
| H3E ———— (Single sideband, full carrier) | | Narrow band FM ———— F3E |
| R3E ———— (Single sideband, reduced carrier) | | |
| J3E ———— (Single sideband, suppressed carrier) | | |
| A3C ———————————— | *Facsimile* | ———————— F3C |
| C3F ———————————— | *Television* | ———————— C3F |

# Necessary bandwidth

The FCC defines necessary bandwidth as follows: "For a given class of emission, the minimum value of the occupied bandwidth sufficient to ensure the transmission of information at the rate and with the quality required for the system employed." In other words, necessary bandwidth describes how much space on the frequency spectrum is required by the radio signal.

The necessary bandwidth is always expressed with four characters (one letter and three numerals). The position of the letter tells you where the decimal point is located. The letters used are as follows:

> H is used for Hz (0.001 – 999 Hz)
> K is used for kHz (1.00 – 999 kHz)
> M is used for MHz (1.00 – 999 MHz)
> G is used for GHz (1.00 – 999 GHz)

*Examples* (notice how the letter occupies the position of the decimal point):

FREQUENCY FCC DESIGNATION

| | |
|---|---|
| 0.002 Hz | H002 |
| 0.2 Hz | H200 |
| 26.3 Hz | 26H3 |
| 3.4 kHz | 3K40 |

| | |
|---|---|
| 6 kHz | 6K00 |
| 14.5 kHz | 14K5 |
| 180.4 kHz | 180K (Rounding off is needed— maximum of four characters) |
| 1.25 MHz | 1M25 |
| 3 MHz | 3M00 |
| 10 MHz | 10M0 |
| 5.75 GHz | 5G75 |

The most common description of an emission type consists of the three-character designation, such as J3E, for single-sideband signals with suppressed carrier. However, you will be expected to be acquainted with the full designation, which also includes the necessary bandwidth. In the case of the J3E, for example, the full designation would be 2K70J3E.

Examples of commonly encountered emission classes:

| Brief designation | Full designation | Definition |
|---|---|---|
| A1A | 100HA1A | Continuous-wave telegraphy (bandwidth is 100 Hz when code is sent at a rate of 25 words per minute) |
| A3E | 6K00A3E | Telephony, double sideband, full carrier amplitude modulated telephony of commercial quality. (*Tip*: A for all or both sideband (formerly designated A3) |
| A3E | 8K00A3E | Telephony, double-sideband, full-carrier amplitude modulated telephony of broadcast quality. (Note the wider bandwidth) |
| C3F | 6M25C3F | TV vestigial sideband. This refers to the lower sideband component of a television signal. The vestigial sideband occupies considerably less bandwidth than the upper sideband (formerly A5 or F5) |
| F1B | 304HF1B | FM telegraphy. Frequency shift keying. Instead of turning the carrier on and off, the carrier remains on all the time. When code characters are sent, the carrier frequency shifts slightly (formerly F1) |
| F3C | 1K98F3C | FM analog facsimile (formerly F4) |
| F3E | 16K0F3E | Narrow-band FM telephony (commercial quality) (formerly F3) |
| F3E | 180KF3E | FM sound broadcasting telephony (Note the wider bandwidth) |
| G3E | 180KG3E | Phase modulation |

| H3E | 3K00H3E | Telephony, single-sideband, full carrier. Carrier must be of a power level between 3 and 6 dB below peak envelope power (formerly was A3H) (*Tip*: H for half of the normally two sidebands with full carrier) |
| --- | --- | --- |
| J3E | 2K70J3E | Telephony, single-sideband, suppressed carrier (SSSC). The carrier must be suppressed to a power level of at least 40 dB below peak envelope power. (*Tip*: J for junior). *Junior*, because the signal occupies a junior space (one-half) on the frequency band. It contains only one sideband, and a 40 dB suppressed carrier (formerly called A3J) |
| R3E | 2K99R3E | Telephony, single-sideband, reduced carrier. The carrier must be at a power level of 18 dB below peak envelope power (*Tip*: R for reduced carrier) (formerly A3A) |

# Frequency ranges

The frequency spectrum is divided into several segments, such as MF, VHF, and UHF. The FCC assigns frequencies within these ranges for use by various services and agencies. An easy way to remember the list of frequency ranges is to remember the phrase: "Vigorously learning more has value, unlike some emotions." Picture an applicant for a position that requires the General Ratiotelephone Operator License. Which activity do you think will have more value—*vigorously learning more* in order to pass the FCC test or displaying *some emotions* such as frustration, anger, and resentment at the FCC field office? You will agree that *vigorously learning more has value, unlike some emotions*. To determine the frequency range, do the following:

1. List the first letters of the phrase "Vigorously learning more has value, unlike some emotions" vertically on a sheet of paper.
2. Number the letters from zero to seven.
3. The frequency band always starts with a 3, followed by a number of zeros. The vertical numbers on your list indicate how many zeros to add to the 3. The 3 with the proper number of zeros gives you the lower limit of the frequency range. Determine the upper limit by adding another zero to the lower frequency limit.
4. The frequencies are in kilohertz (kHz). To convert to megahertz (MHz), move the decimal three places to the left. For example, 3000 kHz is 3 MHz.

**Frequency chart**

0. Vigorously: Very low frequency (VLF)     3 – 30 kHz
(Add no zeros to the 3)

1. Learning:   Low frequency (LF)     30 – 300 kHz
(Add one zero to the 3)

2. More:          Medium frequency (MF)          300 – 3000 kHz
                                                  (Add 2 zeros to the 3)

3. Has:           High frequency (HF)            3000 – 30,000 kHz
                                                  (Add 3 zeros to the 3)

4. Value:         Very high frequency (VHF)      30,000 – 300,000 kHz
                                                  (Add 4 zeros to the 3)

5. Unlike:        Ultra high frequency (UHF)     300,000 – 3,000,000 kHz
                                                  (Add 5 zeros to the 3)

6. Some:          Super high frequency (SHF)     3,000,000 – 30,000,000 kHz
                                                  (Add 6 zeros to the 3)

7. Emotions:      Extremely high frequency (ELF) 30,000,000 – 300,000,000 kHz
                                                  (Add 7 zeros to the 3)

*NOTE:* After you become familiar with the system, all you will need is the frequency designation. For example, if the range of VHF is desired, the V in VHF reminds you of the V in value. Because value is number 4 on the vertical list, add four zeros to the 3 to arrive at 30,000 kHz. This converts to 30 MHz. The upper limit would be 300 MHz.

Note that two additional frequency bands are located below the VLF band. They are voice frequencies (0.3 kHz – 3 kHz) and extremely low frequencies (0.03 kHz – 0.3 kHz).

# Study questions
# Emission types and frequency ranges

1. What is the lower end of the VHF band?

2. The 2182 kHz distress and listening watch frequency is in what frequency band?

3. What is the brief emission designation for amplitude modulation with full carrier and both sidebands?

4. Which emission type is required to have its carrier suppressed by 40 dB?

5. What is the brief emission designation for FM vestigial sideband as used in television?

6. A method of telecommunication where a fixed image is transmitted and converted into a permanent record at the receiving station is called _____.

7. If you walked into a station control point, where would you look to find out what type of emission is being used?

8. What is the designation for a signal with single sideband and full carrier?

9. What is the frequency range for the very low frequency (VLF) band?

10. Which of the following emissions occupies the most space on the frequency band: 6K00A3E, 6M25C3F, 100HA1A, or 304HF1B?

11. Which signal occupies the least amount of space, 3K00H3E or 100HA1A?

# 4
# Direct current

## Conductors versus insulators

Atoms (elements) are the building blocks of all matter. Although there are only slightly more than 100 known elements, they can be combined to form every organic and inorganic compound in existence. The hydrogen atom, the simplest atom, has one positively charged *proton* and one negatively charged *electron*. See Fig. 4-1. Heavier atoms consist of more protons and an equal number of electrons. Heavier atoms also have neutral (without a charge) *neutrons*.

The combination of protons and neutrons form the nucleus of the atom. The nucleus is like our sun. The electrons are like the planets that revolve around the sun. Like the planets vary in their distance from the sun, the orbits of electrons vary in their distance from the nucleus of the atom. Hydrogen, the lightest atom, has only one orbital level (shell), consisting of one electron. Heavier atoms have several orbital levels of electrons. It is the outer shell of orbiting electrons that determines whether an atom is a good conductor or a good insulator. Figure 4-2 illustrates the copper atom.

These outer-shell electrons, sometimes called *valence electrons*, can be either tightly bound or loosely bound. The loosely bound electrons are called *free electrons* because they are free to flow (or drift) from their orbit to the orbit of a nearby atom. Hence, the flow of electrical current is the mass movement of valence electrons in a conductor. For example, if a copper wire is connected to the positive and negative terminals of a battery, a large amount of current will flow through the wire. (Please do not do this. With no resistive load in the fine, the wire will immediately become extremely hot.)

The negative terminal of the battery has an excess of electrons and the positive terminal has a deficiency of electrons. The free electrons in the wire will flow away from the negative terminal and toward the positive terminal, much like water flows

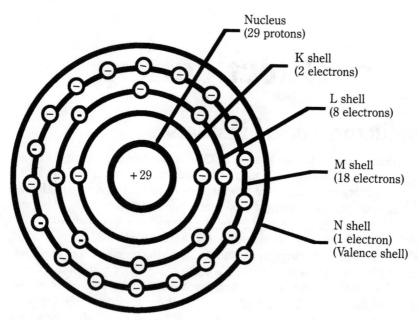

**4-1** Hydrogen Atom.

**4-2** . Copper Atom.

**Table 4-1.**
**Conductors versus insulators.**

| Conductors | Insulators |
|---|---|
| • Loosely bound | • Tightly bound |
| • 1 or 2 electrons | • 7 or 8 electrons |
| Examples: | Examples: |
| • Copper | • Polystyrene |
| • Silver | • Mica |
| • Gold | • Teflon |

Valence electrons determine whether an atom is a
conductor or insulator.

through a hose. Copper is a good conductor because it has many free electrons in its outer shell. If polystyrene is connected to the terminals of a battery, no current will flow because polystyrene has tightly bound electrons in its outer shell. Hence, polystyrene is a good insulator. Some materials are good insulators at power frequencies but poor insulators at radio frequencies. Good radio frequency insulators are mica, quartz, Teflon, and polystyrene. Examples of good conductors are copper, aluminum, and silver. Table 4-1 compares conductors and insulators.

# Voltage

- EMF (electromotive force)
- Difference of potential
- *IR* drop

## Electrical units

- Kilovolts     $10^3$ V     1000 V
- Volts
- Millivolts    $10^{-3}$ V    0.001 V (1mV)
- Microvolts   $10^{-6}$ V    0.000001 V (1 $\mu$V)

Answering FCC questions usually requires the ability to convert from one unit to another. It is therefore, vitally important to learn an easy method of doing so. For example, 0.03 V is equal to how many millivolts? The following explanation will enable you to easily visualize this.

Take a blank sheet of paper. Across the top, list: kilovolts, volts, millivolts, and microvolts. Under each, draw three lines representing the ones, tens and hundreds place values. Then place a decimal point between each set of three lines, as shown below:

| (KV)      | (V)   | (mV)       | ($\mu$V)   |
|-----------|-------|------------|------------|
| Kilovolts | volts | millivolts | microvolts |

$$\underline{\ }\ \underline{\ }\ \underline{\ } \quad . \quad \underline{\ }\ \underline{\ }\ \underline{\ } \quad . \quad \underline{\ }\ \underline{3}\ \underline{0} \quad . \quad \underline{\ }\ \underline{\ }\ \underline{\ }$$

To use the method, simply write your voltage under the appropriate unit category. For example, insert the number 30 under millivolts. To convert to volts, read the number 30 using only the decimal point under *volts*. Notice that 30 mV now reads as 0.03 V. Using the microvolt heading, the number reads as 30,000 $\mu$V. By the same method, you can see that 2500 V is the same as 2.5 kV. or 2,500,000 mV.

| (KV)      | (V)   | (mV)       | ($\mu$V)   |
|-----------|-------|------------|------------|
| Kilovolts | volts | millivolts | microvolts |

$$\underline{\ }\ \underline{\ }\ \underline{2} \quad . \quad \underline{5}\ \underline{0}\ \underline{0} \quad . \quad \underline{\ }\ \underline{\ }\ \underline{\ } \quad . \quad \underline{\ }\ \underline{\ }\ \underline{\ }$$

The measuring device is the *voltmeter*—formed by adding a high-value resistor (multiplier) in series with a basic meter movement. The sensitivity of voltmeters is expressed in ohms per volt and is determined by one of two methods:

- Divide the value of the multiplier resistance by the full-scale voltage value. For example, if the value of the multiplier resistor is 400,000 Ω with a 400-V full-scale reading, calculate as follows:

$$\frac{400,000 \ \Omega}{400} = 1000 \ \Omega/V$$

- Take the reciprocal of the full-scale meter current. A 0 – 1 mA meter would have a sensitivity of 1000 Ω/V.

$$\frac{1}{0.001 \ amp} = 1000 \ \Omega/V$$

# Current

Current (*I*) is the flow or drift of electrons that takes place when a difference of potential (voltage) is applied to a circuit.

## Electrical units

- Ampere
- Milliampere    $10^{-3}$ A    0.001 A (one mA)
- Microampere    $10^{-6}$ A    0.000001 A (one μA)

Use the method described above for units of current:

| A | mA | μA |
|---|----|----|
| Amperes | milliamperes | microampere |

|  |  |  | . |  |  | 4 | . | 5 | 0 | 0 |
|--|--|--|---|--|--|---|---|---|---|---|

By using this method, you can see that 4500 μA is equal to 4.5 mA. And, 0.002 A is the same as 2 mA or 2000 μA.

The measuring device is the *ammeter*—formed by placing a low-value resistor (shunt) in parallel with a basic meter movement. The value of the shunt resistor determines the full-scale current range of the ammeter.

# Resistance

Resistance is the opposition to the flow of electrical current.

## Electrical units

- ohm
- Kilohm    $10^3$ Ω    1000 Ω (one kΩ)
- Megohm    $10^6$ Ω    1,000,000 (one MΩ)

The method described for volts and amperes is useful for units of resistance. For example, 6500 kΩ is the same as 6.5 MΩ, as shown below:

MΩ  kΩ  Ω
Megohm  Kilohm  Ohm

$$\underline{\phantom{6}}\ \underline{\phantom{6}}\ \underline{6}\ .\ \underline{5}\ \underline{0}\ \underline{0}\ .\ \underline{\phantom{6}}\ \underline{\phantom{6}}\ \underline{\phantom{6}}\ \underline{\phantom{6}}$$

The measuring device is the *ohmmeter*.

# Resistors

*Resistors* are circuit components that offer varying degrees of opposition to the flow of current. The amount of opposition depends on the resistor value. A technician can determine the actual value of a resistor by measuring it with an ohmmeter. However, the resistance value is displayed on the resistor by means of a color code. Figure 4-3 illustrates the four color bands. The first three bands relate to the resistance value. The fourth band, or lack of a band, determines the tolerance of the component.

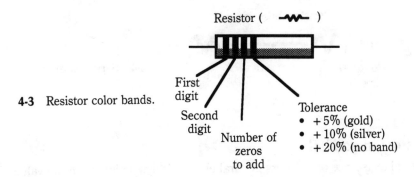

**4-3**  Resistor color bands.

A former student shared a method of remembering the colors. He said, when he was young, he was taught that the colors of the rainbow could be learned by remembering the name, "Roy G. Biv." The man's name relates to the colors, red, orange, yellow, green, blue, indigo, and violet. However, when it comes to the resistor color code, the *I* (indigo) is left out. He went on to say that he remembered that the value of white was nine by thinking of cloud nine. Gray is almost white, hence its value is only eight. Table 4-2 shows how to use the resistor color code.

By using the table, you can see that a 75-Ω resistor with a 5% tolerance is color coded as violet, green, black, and gold. A resistor coded, black, brown, red and silver would be a 100-Ω resistor with 10% tolerance.

## Resistors in series

The total resistance of resistors in series is calculated as follows: See Fig. 4-4.
$$R_{\text{total}} = R_1 + R_2 + R_3 \ldots$$

**Table 4-2. Resistor color code.**

| Memory tip | Color | Digit number | Number of zeros to add |
|---|---|---|---|
| Black Hole | black | 0 | 0 |
| Earth | brown | 1 | add 1 zero |
| R | red | 2 | add 2 zeros |
| O | orange | 3 | add 3 zeros |
| Y | yellow | 4 | add 4 zeros |
| G | green | 5 | add 5 zeros |
| B | blue | 6 | add 6 zeros |
| V | violet | 7 | add 7 zeros |
| Almost white almost 9 | gray | 8 | add 8 zeros |
| Cloud nine | white | 9 | add 9 zeros |

*Roy G. Biv

Resistors in series, add

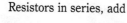

100 Ω                 100 Ω          **4-4**   Series resistors.

$$R_t = 200 \ \Omega$$

## Resistors in parallel

The total resistance of resistors in parallel will always be less than the value of the smallest resistor. Total resistance is calculated as follows when only two resistors are in the circuit.

$$R_{total} = \frac{R_1 \times R_2}{R_1 + R_2}$$

When two resistors of equal value are placed in parallel, the new value will be one half the value of either resistor. For example, if two 100-Ω resistors are placed in parallel, the equivalent resistance would become 50 Ω. See Fig. 4-5. When more than two resistors are in parallel, the following formula is used:

$$R_{total} = \frac{1}{\dfrac{1}{R_1} + \dfrac{1}{R_2} + \dfrac{1}{R_3} + \dfrac{1}{R_4}} \ \cdots$$

## Resistance versus wire size

The resistance varies inversely to the cross-sectional area of the wire conductor. Table 4-3 shows that if the area is doubled, the resistance would be one half. If the

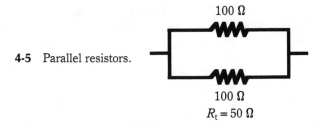

**4-5** Parallel resistors.

100 Ω

100 Ω

$R_t = 50$ Ω

**Table 4-3. Resistance
versus cross-sectional area.**

| Cross-sectional area | Resistance |
|---|---|
| 2X | $1/2$ |
| 3X | $1/3$ |
| 4X | $1/4$ |
| $1/2$X | 2 |

As the cross-sectional area increases, the resistance is reduced.

area is tripled, the resistance would be reduced to one third of the original value. (2 times the area equals one half the resistance; 3 times equals one third; 4 times equals one fourth; etc.). For example, a length of wire has a resistance of 100 Ω. What happens if that wire is replaced with a wire of equal length but with one half the cross-sectional area? When the area is reduced by one half, the resistance is doubled. Figure 4-6 shows that as the cross-sectional area increases, the resistance decreases. The new wire would have a resistance of 200 Ω. For wire diameter, take another step to reduce it to a cross-sectional area problem. The cross-sectional area varies with the square of the diameter of the wire. So if the 100-Ω wire were replaced with a wire of one half the diameter, the cross-sectional area would be one fourth. With an area of one fourth, the resistance would be four times greater or 400 Ω. If the 100-Ω wire is replaced with a wire of the same length but twice the diameter, the new wire would have a resistance of 25 Ω. Be very cautious while

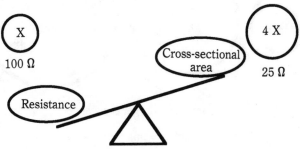

**4-6** Resistance versus cross-sectional area.

reading FCC questions of this nature. Be aware of whether they are asking about changing the cross-sectional area or the diameter of the wire. Carefully study Fig. 4-7.

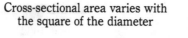

Cross-sectional area varies with
the square of the diameter

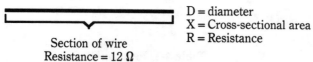

D = diameter
X = Cross-sectional area
R = Resistance

Section of wire
Resistance = 12 Ω

| Diameter of new wire | Cross-sectional area of new wire | Resistance of new wire |
|---|---|---|
| 2 D | 4 X | 1/4R (or 3 Ω) |
| 1/2 D | 1/4 X | 4 R (or 48 Ω) |

**4-7** Resistance versus wire diameter.

## Resistance versus gauge of wire

The lowest gauge number is the largest wire and, therefore, has the lowest resistance. Of wires of 12, 10, and 4 gauge, the 4-gauge wire has the lowest resistance because it has the largest diameter. Lower gauge wire has a higher current-handling capacity. Figure 4-8 illustrates the gauge versus, size, and resistance.

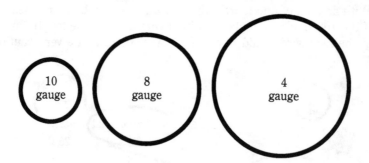

Larger number = larger resistance

Smaller number = smaller resistance

**4-8** Resistance versus wire gauge.

## Power-handling capability

When resistors are placed in parallel, they will handle a greater power dissipation. For example, a 10-W resistance of 100 $\Omega$ is needed. If all that is available are 5-W resistors, two 5-W resistors with a value of 200-$\Omega$ could be used in parallel. As stated above, the value would be one half the individual resistances. The power-handling ability would be 10 W. This holds true only if the value of both resistors is the same. For example, if a 50-$\Omega$, 10-W resistor is placed in parallel with a 100-$\Omega$, 10-W resistor, the total power-handling capability would be about 15 W. This is true because if the 100-$\Omega$ resistor were made to dissipate its full 10 W, the 50-$\Omega$ resistor would exceed its capacity and burn out.

# Ohm's law

Ohm's law is the mathematical expression of the relationship between:

voltage ($E$)
current ($I$)
resistance ($R$)
power ($P$)

There are twelve formulas. If you know three of them, you can transpose and determine the other formulas.

| | | |
|---|---|---|
| 1. $I = \dfrac{E}{R}$ | 2. $E = I \times R$ | 3. $R = \dfrac{E}{I}$ |
| 4. $P = I \times E$ | 5. $I = \dfrac{P}{E}$ | 6. $E = \dfrac{P}{I}$ |
| 7. $P = I^2 R$ | 8. $I = \sqrt{\dfrac{P}{R}}$ | 9. $R = \dfrac{P}{I^2}$ |
| 10. $P = \dfrac{E^2}{R}$ | 11. $E = \sqrt{P \times R}$ | 12. $R = \dfrac{E^2}{P}$ |

You should be familiar with the above combinations, as well as applications that require their use. For example, formula 7 can be used in the direct method of transmitter power calculation. Formula 1 through 3 are most frequently used, so review them accordingly.

The circles on Fig. 4-9 might help you to remember Ohm's Law. Simply place your finger over the electrical unit that you want to determine. For example, if you know the resistance and voltage, determine the current flow by covering the $I$ in

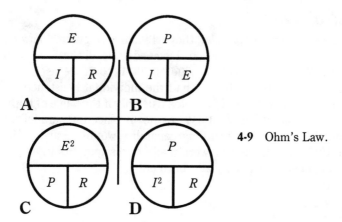

**4-9**   Ohm's Law.

Fig. 4-9a with your finger. The remaining part of the circle tells you to divide the voltage ($E$) by the resistance ($R$). In Fig. 4-9b, you can determine the power in a circuit by multiplying the current by the voltage. You will notice in Fig. 4-9 (c and d) that the square is removed from the $E$ and the $I$ by placing a square root sign over the remaining combination. This is shown in formulas 8 and 11 above.

# Lead-acid batteries

The widely used lead-acid battery will be discussed because of its use in portable and mobile installations.

**Structure**   The lead-acid battery consists of:
- A solid lead peroxide anode (+)
- A solid sponge lead cathode (−)
- A sulfuric acid electrolyte solution

## Condition of battery

Use a hydrometer to determine the level of charge. The hydrometer measures the specific gravity of the electrolyte solution. The electrolyte weighs more in a fully charged battery because there is more acid in solution. As the battery discharges, the sulfuric acid combines with the plates. This leaves more water than acid. Because the water weighs less than the sulfuric acid, the specific gravity is less when the battery is discharged. The specific gravity of distilled water is 1.000 and the specific gravity of a fully charged battery is about 1.300.

$$\text{Specific gravity} = \frac{\text{weight of electrolyte}}{\text{weight of water}}$$

## Long-term storage

If a battery is stored for a long duration, the electrolyte should be drained and replaced with distilled water.

## Care of batteries

- Protect the battery from extreme temperature variations because temperature variations shorten its life.
- Keep electrolyte level slightly above the top of the plates.
  - ~ Add water when level is low. Clean or distilled water should be used to prevent plate deterioration.
  - ~ The only time acid needs to be added is when it spins out. The acid does not evaporate like water does.
- Check specific gravity with a hydrometer to determine charge.
- When charges at too high a rate, excess gasses will be released. This produces an explosive condition if adequate ventilation is not present, and the plates tend to deteriorate. A slower charge is preferable.
- A *trickle* (slow) charge should be placed on a battery that is not being used. This maintains the change.
- Maintain good cable connections.

## Battery capacity ratings

One method of rating batteries is by the number of *ampere-hours* (Ah).

*Example 1*   A battery that is rated 60 Ah can be discharged at a rate of 10 A per hour for 6 hours. If less current is drawn, the life of the battery is extended.

$$\frac{60 \text{ Ah}}{6 \text{ h}} \text{ 10 A for 6 h}$$

*Example 2*   A 12-V battery that is rated at 100 Ah on an 8-hour basis would supply what current to the load?

$$\frac{100 \text{ Ah}}{8 \text{ h}} = 12.5 \text{ A for 8 h}$$

*Example 3*   A transmitter requires 350 W of power and a receiver requires 50 W of power from a 12-V 50 amp-hour battery source. How long would it take to discharge the battery?

1. First, the amount of current must be determined as follows:

$$I = \frac{P}{E}$$

Where

$I$ = current, in amperes
$P$ = power, in watts
$E$ = voltage, in volts

$$\text{therefore, } I = \frac{400 \text{ W}}{12 \text{ V}} = 33.3 \text{ A}$$

2. Then calculate the number of hours required to discharge the battery:

$$\frac{50 \text{ Ah}}{\text{hours}} = 33.3 \text{ A}$$

or

$$\text{hours} = \frac{50 \text{ Ah}}{33.3 \text{ A}} = 1.5 \text{ h}$$

Table 4-4 summarizes the steps to determine how long it takes for a battery to fully discharge. It would be wise for you to try various combinations and become very familiar with this type of problem.

### Table 4-4. Ampere-hour calculations.

Question:  How long to discharge battery
Knowns:   1) Ampere-hour rating of battery
          2) Battery voltage
          3) Amount of power drain on battery
       ✔ 1. Calculate current drain.

$$I = \frac{P}{E}$$

✔ 2. Divide ampere-hour rating by the current drain.

$$\frac{\text{Ah}}{\text{A}}$$

✔ 3. If decimal is formed — calculate minutes.

$$60 \times .4 = 24 \text{ min}$$
$$60 \times .6 = 36 \text{ min}$$
$$60 \times .8 = 48 \text{ min, etc.}$$
$$4.8 \text{ h} = 4 \text{ h } 48 \text{ min}$$

# Study questions
# Direct current

1. Electrical current is a _____ of electrons that takes place when a _____ is applied to a circuit.

2. Voltage can also be called _____.

3. The sensitivity of a voltmeter is expressed in _____.

4. The _____ can be used to measure electrical resistance.

5. The resistance of a wire varies inversely with the _____ of the conductor.

6. If the cross-sectional area of a conductor is tripled, the resistance will be reduced to _____.

7. As the gauge of wire is decreased, the resistance is _____.

8. $R_1 + R_2 + R_3$ is the method of calculating the resistance of resistors connected in _____.

9. A conductor has many _____ in the outer shells of its atoms.

10. Examples of good insulators at radio frequencies are: _____

11. How is power found when current and resistance are known?

12. The charge of a battery can be checked with a _____, which measures the specific gravity of the electrolyte.

13. How should a battery be stored for a long duration?

14. How is the capacity of a battery rated?

15. What is the maximum power dissipation of two 10-W resistors connected in parallel, where one is 50 Ω and the other is 100 Ω?

16. The voltage across a variable resistor is doubled. According to Ohm's law, how must its resistance be adjusted in order to maintain the same amount of power dissipation?

17. What is the most significant factor in determining if a material is a conductor or an insulator?

18. A 12-in length of copper wire, having a resistance of 100 Ω is replaced with a 12-in section of wire with four times the cross-sectional area. What is the resistance of the new piece of wire?

19. If a resistor is color coded, red, orange, yellow, and gold, what is the resistance?

20. A transmitter requires 300 W of power, a receiver uses 100 W and miscellaneous electrical devices use an additional 80 W of power. The power supply is a 12-V battery with an 80 Ah rating. How long will it take to fully discharge the battery, if all the above devices are operating?

21. A voltage of 0.02 kV is equal to how many volts?

# 5
# Alternating current

While direct current (dc) maintains a constant amplitude and direction of flow, alternating current (ac) constantly changes amplitude and periodically changes in direction of flow. This resultant ac waveform is called a *sine wave*. The chapter on motors and generators explains how the sine wave is produced.

## Frequency of a sine wave

One complete sine wave is one cycle. One complete cycle in one second is one cycle per second (cps). The term *hertz* (Hz) has been substituted for cycles per second in recent years. 1000 cycles occurring in a second are 1000 cps or 1000 Hz or 1 kHz. The following is a summary of the units used in measuring frequency of sine waves:

- Hertz (Hz)         1 cps                              (1 Hz)
- Kilohertz (kHz)    $10^3$ Hz    1000 Hz         (1 kHz)
- Megahertz (MHz)   $10^6$ Hz    1,000,000 Hz     (1 MHz)
- Gigahertz (GHz)    $10^9$ Hz    1,000,000,000 Hz   (1 GHz)

Use the following conversion method:

| GHz | MHz | kHz | Hz |
|-----|-----|-----|-----|
| Gigahertz | Megahertz | Kilohertz | Hertz |

  ___ ___ ___ . ___ ___ 2 . 1 8 2 . ___ ___ ___

Note that 2182 kHz is the same as 2.182 MHz.

# Period of a sine wave

The period is the time required for a sine wave to progress through one complete cycle (360° of movement). The period is also called the *time* of a sine wave. Mathematically, it is represented by:

$$\text{Time} = \frac{1}{\text{Frequency}}$$

For example: Determine the time, or period, of a 1-MHz sine wave. That is, how long does it take to complete one full cycle?

$$\text{Time} = \frac{1}{1,000,000} = 0.000001 \text{ s}$$

When the decimal point is moved six places to the right, the answer becomes 1 $\mu$s. The relationship between *time* and *frequency* is as follows:

$$\text{Time} = \frac{1}{\text{Frequency}}$$

$$\text{Frequency} = \frac{1}{\text{Time}}$$

# Amplitude of a sine wave

There are several ways of expressing the amplitude of a sine wave:

**Peak amplitude**   The peak amplitude of a *sinusoidal wave* (sine wave) is determined by the maximum deviation in the positive or negative direction. Since a sine wave is symmetrical, the positive peak will be the same value as the negative peak.

**Peak-to-peak amplitude**   Peak-to-peak amplitude is determined by the distance from the positive peak to the negative peak. If the sine wave is symmetrical, where the positive peak and negative peaks are equal, simply double the peak value to get peak-to-peak amplitude.

*Example*   If the positive peak value is 12 V and the negative peak value is 2 V, the peak-to-peak value is 14 V.

*Example*   If the positive and negative peaks are both equal to 10 V, the peak-to-peak value is 20 V.

**Root-mean-square amplitude (rms)**   The rms amplitude is the most commonly used method of describing the amplitude of a sine wave. Most ac meters measure in rms values. The rms value is also called the *effective* value. A light bulb connected to 100 Vdc will be equally bright when connected to 100 V rms. However, if 100 V peak value is connected to the bulb, its intensity would be reduced. This is because the 100 Vdc remains as the full 100-V level constantly. The 100-V peak sine wave only reaches 100 V on the peaks. Therefore, the effective value of the ac wave is less than the value of the dc source. The *effective (rms)* value is equal to 70.7% of the peak, or 0.707 times the peak value. Determine peak by multiplying the effective value by 1.414.

*Example*   What is the effective value of 100 V peak?

$$100 \times .707 = 70.7 \text{ V}$$

Therefore, 100 V peak would be like 70.7 Vdc.

*Example*   How much peak ac voltage would be required to produce the same intensity of a light bulb that 100 Vdc would produce?

$$100 \times 1.414 = 141.4 \text{ V}$$

Therefore, 141.4 Vac peak is equivalent to 100 Vdc. That is, they would both cause a bulb to light with equal intensity. Another way of saying it is that they both cause the same amount of power dissipation through the filament of the bulb.

**Average amplitude**   Because the ac sine wave constantly changes in amplitude, an average can be determined. Because the average of the entire cycle would be zero, the average is taken from a half cycle of the sine wave. The average value is equal to 0.637 times the peak value.

In summary, keep in mind that most ac meters are calibrated in rms values. It is important to consider peak values while working with diodes. Table 5-1 summarizes the relationships between various values of sine wave amplitude. Figure 5-1 illustrates a sine wave that has a 70.7-V rms value.

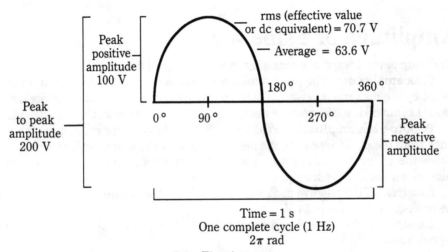

Time = 1 s
One complete cycle (1 Hz)
$2\pi$ rad

**5-1**   The sine wave.

**Table 5-1.**
**The relationship between values of sine wave amplitude.**

| Peak-to-peak value | Peak value | rms (effective) value | Average value |
|---|---|---|---|
| 2 × peak | 1.414 × rms | 0.707 × peak | 0.636 × peak |
| 2.828 × rms | 1.58 × avg. | 1.11 × avg. | 0.9 × rms |
| 200 V | 100 V | 70.7 V | 63.6 V |
| 339.4 V | 169.7 V | 120 V | 108 V |

Remember that 70.7 V rms in an ac circuit and 70.7 V in a dc circuit will produce the same amount of power dissipation when applied across the same value of resistance. This is why rms is called the *dc equivalent*.

# The square wave

The square wave is another form of alternating current. Like the sine wave, it travels through positive and negative cycles. The sine wave gradually changes amplitude as it moves through its cycle. The square wave maintains a constant positive amplitude, then abruptly changes to a constant negative amplitude. Its frequency is determined by the number of times per second the wave progresses through positive and negative cycles. Figure 5-2 illustrates the difference between a square wave and a sine wave. The square wave occupies a wider bandwidth than a pure sine wave because it contains numerous harmonics. If a square wave is analyzed, you will find a fundamental sine wave with numerous odd harmonics superimposed. Although a perfect square wave contains an infinite number of odd harmonics, five to seven odd harmonics produce a respectable square wave. To determine the value of sine wave harmonics necessary to build a square wave, simply multiply the fundamental sine wave frequency by 3, 5, 7, 9, 11, 13, etc. For example, if the fundamental sine wave is 25 kHz, the following can be added to produce a square wave:

$$25 \text{ KHz} \times 1 = 25 \text{ kHz}$$
$$25 \text{ KHz} \times 3 = 75 \text{ kHz}$$
$$25 \text{ KHz} \times 5 = 125 \text{ kHz}$$
$$25 \text{ KHz} \times 7 = 175 \text{ kHz}$$
$$25 \text{ KHz} \times 9 = 225 \text{ kHz}$$

Note that only odd harmonics are included. If even harmonics are used, a triangular wave will be formed.

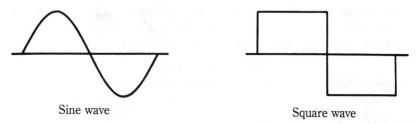

Sine wave          Square wave

**5-2**   Sine wave versus square wave.

# ac circuit components—the capacitor

A *capacitor* is an electrical device having two conductive plates separated by an insulator (*dielectric*). When voltage is applied to a capacitor, an electrical charge is developed. This electrostatic charge remains indefinitely unless purposely discharged. The capacitor produces an opposition to the flow of current in an ac circuit. This opposition or resistance is called *capacitive reactance ($X_C$)*. The capacity of a capacitor can be increased by adding plates, increasing the area of the plates, decreasing the distance between the plates, and increasing the dielectric constant of the insulator between the plates.

## Electrical units

Farad (F)
Microfarad ($\mu$F)    $10^{-6}$F    0.000,001 F (1 $\mu$F)
Picofarad (pFd)    $10^{-12}$F    0.000,000,000,001 F (1 pF)

**Capacitors in series**    When capacitors are connected in series, they act like resistors connected in parallel. That is, an equivalent value is produced that is less than the value of the lowest value of capacitance in the circuit. When capacitors are placed in series, the value of capacitance decreases. The calculation of total capacitance is made as follows:

$$C_{total} = \frac{1}{\dfrac{1}{C_1} + \dfrac{1}{C_2} + \dfrac{1}{C_3} \cdots}$$

*Example*    Calculate the series capacitance in a circuit where:
$C_1 = 10 \ \mu$F
$C_2 = 20 \ \mu$F
$C_3 = 30 \ \mu$F

$$C_{total} = \frac{1}{\dfrac{1}{10} + \dfrac{1}{20} + \dfrac{1}{30}}$$

$$= \frac{1}{0.1 + 0.05 + 0.03} = 5.55 \ \mu F$$

**Capacitors in parallel**    When capacitors are connected in parallel, they act like resistors connected in series. They are simply added together to form a larger capacitance. Parallel capacitance is calculated as follows:

$$C_{total} = C_1 + C_2 + C_3 \ldots$$

*Example*    Calculate the parallel capacitance in a circuit where:
$C_1 = 10 \ \mu$F
$C_2 = 20 \ \mu$F
$C_3 = 30 \ \mu$F
$C_{total} = 10 + 20 + 30 = 60 \ \mu$F

**Capacitive reactance ($X_C$)**    Capacitive reactance is the resistance or opposition to flow of alternating current that a capacitor produces. It is calculated as follows:

$$X_C = \frac{1}{6.28 \times f \times C}$$

where

6.28 $= 2 \times \pi$

f = frequency in Hertz

C = capacitance in farads

*Example*  Calculate the capacitive reactance where:

$f$ = 1000 Hz (1 kHz)

$C$ = 50 $\mu$F

*NOTE:* When the capacitance is in microfarads, the formula can be changed as follows to simplify calculations:

$$X_C = \frac{10^6}{6.28 \times 1000 \times 50} = 3.185 \ \Omega$$

## Capacitive reactance versus frequency

According to the capacitive reactance formula, capacitive reactance is inversely proportional to the frequency. That is, as the frequency increases, the capacitor reactance decreases. This characteristic of capacitors allows the selective attenuation of lower frequencies. In the circuit shown in Fig. 5-3, higher frequencies are allowed to pass through the capacitor. Direct current and low frequencies are blocked by the capacitor and forced to go through the resistance, where they are attenuated. If the input frequency is steadily increased, while maintaining a constant amplitude, the output of the circuit will steadily increase. This is because the higher frequencies receive less attenuation through the capacitor, because of the lower reactance. The lower the frequency, the greater the attenuation through the resistor. This principle is used in the pre-emphasis circuit of the FM transmitter. The higher voice frequencies are amplified to a greater extent than the lower voice frequencies. In the FM receiver, a de-emphasis circuit does the opposite, bringing the voice waveform back to normal. This process increases the signal-to-noise ratio. Because capacitors offer a low resistance path for high frequencies, they are useful in dc power supply circuits. In this case, the capacitor is used to bypass ac signals to ground.

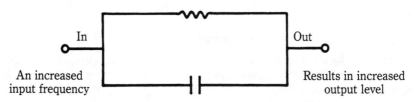

**5-3**  Capacitive reactance versus frequency.

# ac circuit components—the inductor

The inductor is an electrical device consisting of a coil of wire. As alternating current passes through the coil, an expanding and contracting magnetic field is produced. The strength of the field is proportional to the square of the number of turns. If you replace an air-core inductor with an iron core, the inductance will increase. The inductor produces an opposition to the flow of current in an ac circuit. This opposition is called *inductive reactance*.

## Electrical units

Henry (H)
Millihenry    $10^{-3}$ H    0.001 H (1 mH)
Microhenry   $10^{-6}$ H    0.000,001 H (1 $\mu$H)

**Inductors in series**    When inductors are connected in series, they act like resistors connected in series. That is, their values are simply added, producing a larger value of inductance. Use the following formula to calculate inductors connected in series:

$$L_{total} = L_1 + L_2 + L_3 \ldots.$$

**Inductors in parallel**    Total inductance is calculated in the same manner as total resistance in both series and parallel circuits. Therefore, calculate inductors in parallel as follows:

$$L_{total} = \frac{L_1 \times L_2}{L_1 + L_2} \text{(for 2 inductors)}$$

$$L_{total} = \frac{1}{\dfrac{1}{L_1} + \dfrac{1}{L_2} + \dfrac{1}{L_3} \ldots.} \text{ (for any number of inductors)}$$

**Inductive reactance** ($X_L$)    Inductive reactance is a resistance or opposition to the flow of alternating current. It is produced by an inductor. Inductive reactance is calculated as follows:

$$X_L = 6.28 \times f \times l$$

where

6.28 = $2 \times \pi$
$f$ = frequency in hertz
$l$ = inductance in henries

## Inductive reactance versus frequency

According to its formula, inductive reactance is directly proportional to frequency. That is, as the frequency increases, the reactance also increases. This is in direct contrast to the characteristic of the capacitor. As application of inductors can restrict ac from entering a dc power supply. For example, an RF choke can be connected between the collector of a transistor to the supply voltage. The choke allows the dc from the power supply to pass, but prevents the ac signals from the transistor circuit from entering the power supply. Table 5-2 summarizes the major differences between inductors and capacitors. Notice that, in a broad sense, the capacitor acts like a high-pass filter, and the inductor acts like a low-pass filter.

**Table 5-2. Characteristics
of inductors and capacitors.**

| Capacitor (high-pass) | Inductor (low-pass) |
| --- | --- |
| Blocks dc | Passes dc |
| Attenuates low frequencies | Passes low frequencies |
| Passes high frequencies | Attenuates high frequencies |

# Phase relationships in ac circuits

Voltage and current are *in phase* or *in step* in a purely resistive circuit. That is, the voltage and current have their maximum and minimum values at the same points in time. As a component of reactance is introduced from a capacitor or inductor, the phase difference steadily increases. In other words, the voltage and current reach their maximum values at different points in time. When the resistance and reactance are equal, the phase difference is 45°. It continues to increase as more reactance is introduced. Finally, when the circuit contains pure reactance and no resistance, the phase difference between the voltage and the current is 90°. To determine which is leading and which is lagging, one can remember the phrase "ELI the ICE man." That is, in an inductive circuit (L), the voltage (E) leads the current (I). In a capacitive circuit (C), the current (I) leads the voltage (E). To remember that, in a resistive circuit the voltage and current are in phase, remember the Ohm's law formula for resistance. In that formula, notice that neither the voltage (E) nor the current (I) are leading. They are both lined up and in phase.

## Purely resistive circuit

An ac circuit containing only resistance has no phase differences. That is, the voltage (E) and current (I) in the circuit are *in phase*. See Fig. 5-4.

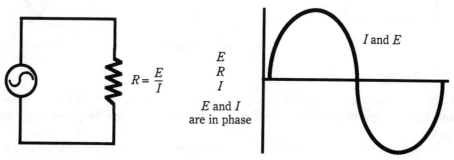

$$R = \frac{E}{I}$$

$$\begin{array}{c} E \\ R \\ I \end{array}$$

*E* and *I*
are in phase

*I* and *E*

**5-4**   Purely resistive circuit.

## Purely inductive circuit

In a purely inductive circuit, ELI is used to remember that the voltage ($E$) leads the current ($I$) by 90°. See Fig. 5-5.

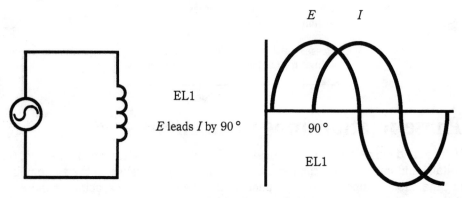

ELI

$E$ leads $I$ by 90°

**5-5** Purely inductive circuit.

## Purely capacitive circuit

In a circuit containing only capacitance, ICE is used to remember that the current ($I$) leads the voltage ($E$) by 90°. See Fig. 5-6.

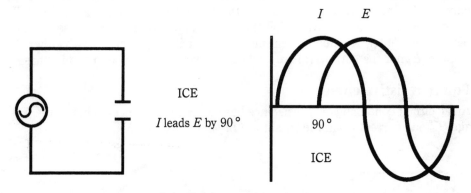

ICE

$I$ leads $E$ by 90°

**5-6** Purely capacitive circuit.

## RLC circuits

In circuits containing a combination of resistance ($R$), inductance ($L$) and capacitance ($C$), a couple of additional steps are required to determine the phase angle as follows:

1. Determine the difference between the capacitive reactance and the inductive reactance. They tend to cancel each other out, ohm for ohm. In the example shown in Fig. 5-7, subtract 40 $\Omega$ from 52 $\Omega$ to arrive at 12 $\Omega$.

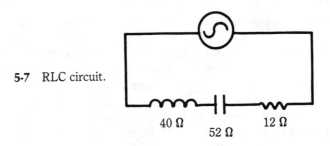

**5-7** RLC circuit.

40 Ω    12 Ω

52 Ω

2. Determine the phase angle by the following formula:

$$\text{Tangent of phase angle} = \frac{X}{R}$$

where

$X$ = reactance
$R$ = resistance

Therefore

$$\frac{12}{12} = 1$$

The inverse tangent of 1 is 45°. Therefore the phase angle in the circuit is 45°.

Remember that when the resistance and the reactance are equal, the phase angle is 45°. In the above case, the resultant reactance is equal to the resistance since the cancellation takes place.

*NOTE:* Trig tables are not allowed in the exam. Calculators with trig functions are allowed.

# Impedance

The components in an ac circuit offer *reactance* (resistance). The capacitor offers capacitive reactance. The inductor offers inductive reactance. And the resistor offers pure resistance. The total value of resistance is called *impedance*. The total reactance is determined by subtracting the capacitive reactance from the inductive reactance. These two offset one another. When they are equal, they cancel, leaving only pure resistance in the circuit. This condition is called *resonance*. The impedance in a series RLC circuit is calculated as follows:

1. Calculate capacitive and inductive reactance according to their formulas, unless given in the problem.
2. Determine total reactance ($X$) by subtracting the capacitive reactance from the inductive reactance ($X_L - X_C$).
3. Calculate impedance with the following formula:

$$\text{Impedance } (Z) = \sqrt{R^2 + X^2}$$

# Time constant

When a capacitor is connected directly across a battery, it charges immediately up to the supply voltage. If a resistance is connected in series with the capacitor, the capacitor charges at a slower rate because it charges through the resistance. (Remember that resistance is defined as the opposition to the flow of electric current.) The time constant in a series resistance-capacitance (RC) circuit is the time required for the capacitor to charge up (through a resistance) to 63.2% (0.632) of the supply voltage. Please note the series configuration in Fig. 5-8a. Fig. 5-8b illustrates that the capacitor charges up to 63.2 percent of the supply voltage after one time constant (TC). It also shows that, for all practical purposes, it take five time constants to fully charge the capacitor.

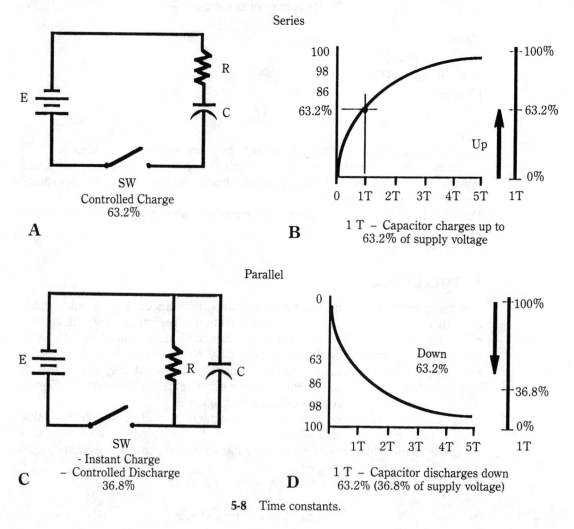

**5-8**   Time constants.

Note that in Fig. 5-8a, the charging begins when the switch is closed. However, in Fig. 5-8c, the capacitor immediately charges to the supply voltage. Notice that, ignoring the resistor, that the capacitor is connected directly across the battery. So in the parallel configuration, the charge is instantaneous. When the switch is then opened, a controlled discharge takes place through the resistor. But in this case, the first time constant is the time required for the capacitor to discharge down 63.2% from the supply voltage. Careful study of Fig. 5-8d will show you that when the capacitor discharges down 63.2%, it is at 36.8% of the supply voltage.

Become very familiar with both circuits and both graphs. Table 5-3 summarizes various time constant definitions.

**Table 5-3. Charging versus discharging time constants for capacitors and inductors.**

| R C<br>T I M E<br>C O N S T A N T | |
|---|---|
| | **Charging capacitor**<br>Time required to charge up to 63.2%<br>of supply voltage. |
| | **Discharging capacitor**<br>Time required to discharge down<br>63.2% from supply voltage to a<br>value of 36.8% of supply voltage. |

| R L<br>T I M E<br>C O N S T A N T | |
|---|---|
| | **Charging inductor**<br>Time required for current to build<br>up to 63.2% of its maximum value |
| | **Discharging inductor**<br>Time required for current to decay<br>to 63.2% of its maximum value. |

*Example* Determine the time constant in Fig. 5-8a:

$$TC = R \times C$$

Where

$TC$ = Time constant
$C$ = 0.01 $\mu$F
$R$ = 2 M$\Omega$

$$TC = 2 \times 10^6 \times 0.01 \times 10^{-6} = 0.02 \text{ s}$$

Therefore, in 0.02 s, the capacitor will charge to 63.2% of the supply voltage. Because the supply voltage is 12 V, the capacitor will charge up to $12 \times 0.632$ or 7.58 V in one time constant of 0.02 s.

*Example* How long will it take the capacitor to charge up to 12 V? Because one time constant ($TC$) is equal to 0.02 seconds, and 5 time constants are required to charge the capacitor up to the supply voltage, the answer is:

$$0.02 \times 5 = 0.1 \text{ s}$$

## Time constant in an RL (resistance-inductance) circuit

The time constant in an RL circuit is the time required for the current in a coil to build up to 63.2% of its maximum value. It takes five time constants to reach the maximum current level.

# Additional practice (capacitance)

A technician must understand how to calculate total capacitance in a series-parallel circuit. Figure 5-9 illustrates the kind of circuit you might want to see on your FCC test. In this case, each capacitor has value of 8 $\mu$F. The following steps show how to determine the total capacitance of the circuit.

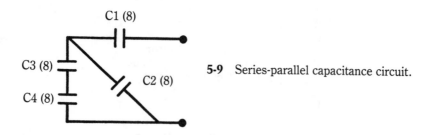

**5-9** Series-parallel capacitance circuit.

1. When the circuit is stretched out, as shown in Fig. 5-10, you can see more easily, which capacitors are in series and which ones are in parallel. Capacitors C4 and C3 are in series. Calculate equi as follows:

$$C_4 + C_3 = \frac{C_4 \times C_3}{C_4 + C_3} = \frac{8 \times 8}{8 + 8} = \frac{64}{16} = 4 \ \mu F$$

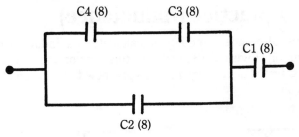

**5-10** Equivalent circuit 1.

2. Figure 5-11 shows the equivalent circuit. Notice that there are now only three capacitors in the circuit. $C_3$ and $C_4$ have been combined to form an equivalent capacitance of 4 $\mu$F.
3. In Fig. 5-11, the 4 and 8 $\mu$F capacitors are in parallel. Because capacitors connected in parallel add, they form an equivalent capacitance of $8 + 4 = 12$ $\mu$F.

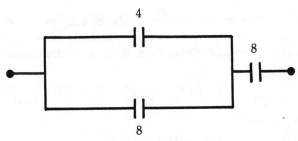

**5-11** Equivalent circuit 2.

4. Figure 5-12 shows the new equivalent circuit. Calculate equivalent capacitance of these final two capacitors as follows:

$$\frac{12 \times 8}{12 + 8} = \frac{96}{20} = 4.8 \ \mu\text{F}$$

Thus, 4.8 $\mu$F is the total capacitance of the complex circuit of Fig. 5-9. Notice that the total capacitance is less than the value of each of the individual capacitors. This is a good rule of thumb to remember. That is, if you see a circuit of equal-value capacitors, the total capacitance will always be less than that value. If the capacitors are 10 $\mu$F, the total capacitance will be less than 10 $\mu$F. If the capacitors are all 2 $\mu$F, any test answers of 2 $\mu$F or higher can immediately be eliminated. Try solving Fig. 5-9 with 2 $\mu$F capacitors, for additional practice.

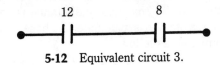

**5-12** Equivalent circuit 3.

# Additional practice (inductance)

A technician must also understand how to calculate total inductance in a complex series-parallel circuit. Figure 5-13 illustrates this type of circuit you can expect to find on your FCC test. The following steps show how to determine the total inductance of the circuit.

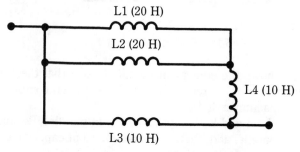

L1 (20 H)

L2 (20 H)

L4 (10 H)

L3 (10 H)

**5-13**   Series-parallel inductance circuit.

1. Figure 5-14 shows the same circuit stretched out. A complex circuit should be redrawn in order to see more clearly which components are in series versus parallel. L1 and L2 are in parallel. Their equivalent inductance is calculated as follows:

$$\frac{L_1 \times L_2}{L_1 + L_2} = \frac{20 \times 20}{20 + 20} = \frac{400}{40} = 10 \text{ H}$$

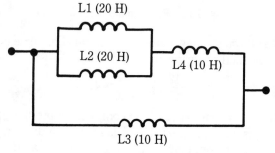

L1 (20 H)

L2 (20 H)

L4 (10 H)

L3 (10 H)

**5-14**   Equivalent circuit 1.

2. When L1 and L2 of Fig. 5-14 are combined, a new equivalent circuit (Fig. 5-15) is formed. This circuit has L3 and L4 from Fig. 5-14 plus the new 10 H inductance from the combination of L1 and L2.
3. Figure 5-16 shows how the circuit looks when the two series 10 H inductors are added together. Notice that total inductance is calculated in the same manner as total resistance is calculated. These two remaining parallel inductors are combined as follows:

$$\frac{20 \times 10}{20 + 10} = \frac{200}{30} = 6.66 \text{ H}$$

Thus, 6.66 H is the total inductance of Fig. 5-13. Notice that the total is smaller than the lowest value of inductance. For additional practice, calculate the total inductance of Fig. 5-13, with inductance values of one half the values shown in that diagram.

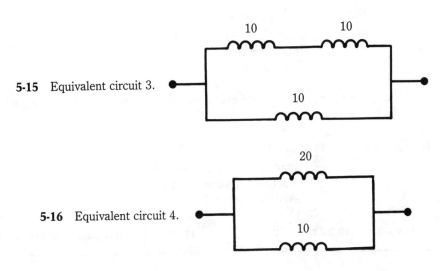

**5-15**   Equivalent circuit 3.

**5-16**   Equivalent circuit 4.

# Study questions
# Alternating current

1. What is the relationship between the frequency and period of a sine wave?

2. The period of a sine wave is the time required for a sine wave to progress through _____ °.

3. Most meters are calibrated to measure _____ amplitude.

4. What is the capacitance of three 100-$\mu$F capacitors connected in series?

5. What is the capacitance of three 200-$\mu$F capacitors connected in parallel?

6. As the frequency is increased, the capacitive reactance _____.

7. If a signal is passed through a circuit consisting of a resistor and capacitor in parallel, what would happen at the output if the input frequency is steadily increased?

8. Inductors in series are like resistors in _____.

9. What is the inductance of two 25-$\mu$H coils connected in parallel?

10. In a purely resistive circuit, the phase angle is _____.

11. In a purely capacitive circuit, the _____ leads _____
the by _____°.

12. What is the phase angle in a circuit where:

> Capacitive reactance is 70 Ω
> Inductive reactance is 30 Ω
> Resistance is 40 Ω

13. What is the phase angle in the above circuit if the resistance is changed to 28 Ω?

14. What is the phase angle in a circuit where:

> Capacitive reactance is 100 Ω
> Inductive resistance is 100 Ω
> Resistance is 35 Ω

15. The time constant is the time (in seconds) required for a capacitor to charge up to _____ % of the supply _____.

16. A circuit has a 100-Ω resistor and a 200-μF capacitor.
a. What is the time constant?
b. How many time constants are required for the capacitor to charge up to the supply voltage?
c. How long would the supply voltage have to be applied in order for the capacitor to charge up to the supply voltage?

17. If the air core of a choke coil is replaced with an iron core, what would happen?

18. What is the impedance of a series circuit, where the resistor is 1 MΩ, the capacitor is 0.001 μF, and $2\pi F$ is 1000?

19. The frequency 2182 kHz is equal to how many gigahertz?

20. If a capacitor is connected directly across a battery, how many time constants are required for it to fully charge?

21. If a resistor and capacitor and battery are all connected in parallel, how many time constants are required to fully charge the capacitor?

22. When a capacitor discharges through a resistor, at what point is the charge equal to 36.8% of its full charge?

# 6
# Transformers

## Transformer operation

Transformer operation is possible because of the following three principles:

- When current flows through a conductor, a magnetic field is generated around that conductor. If the current flowing through the conductor is alternating current, the magnetic field will be alternating as well. That is, the magnetic field will expand and contract in accordance with the frequency of the alternating current. By arranging the conductor in the form of a coil, the strength of the field will be increased proportionally to the square of the number of turns.
- When a conductor is placed close to an alternating magnetic field, a voltage will be *induced* in that conductor, even though there is no physical contact between the two. If this conductor is arranged into a coil, the induced voltage will be larger.
- When two coils are placed in close proximity, coil 1 will induce a voltage in coil 2. They are said to be magnetically coupled. This is called *transformer action*.

### Step-up, step-down

Step-up and step-down refer to the *voltage* relationship of the primary-to-secondary windings of the transformer. See Fig. 6-1.

**Step-up transformer**  This type has more turns in the secondary. A small voltage on the primary will produce a larger voltage on the secondary.

**Step-down transformer**  This type has more turns on the primary. A large voltage on the primary will produce a smaller voltage on the secondary.

**Isolation transformer**  This type has equal number of turns on the primary and secondary.

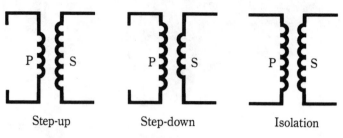

Step-up                Step-down                Isolation

**6-1**   Transformers.

# Relationships between primary and secondary

Primary-to-secondary relationships are summarized by the following formula:

$$\frac{N_p}{N_s} = \frac{E_p}{E_s} = \frac{I_s}{I_p}$$

where

$I$ = current
$E$ = voltage
$N$ = number of turns
$p$ = primary
$s$ = secondary

**Example 1**   Calculate the induced voltage in the secondary of the transformer in Fig. 6-2 where:

There are 400 turns on the primary
There are 2400 turns on the secondary
Voltage on the primary is 110 V
R = 33 kΩ

$$\frac{N_p}{N_s} = \frac{E_p}{E_s}\frac{400}{2400} = \frac{110}{E_s}\frac{(2400)(110)}{400} = E_s = 600 \text{ V}$$

**Example 2**   Calculate the current flow in the secondary circuit of Fig. 6-2.

$$I = \frac{E}{R} = \frac{660}{33,000} = 0.02 \text{ A (20 mA)}$$

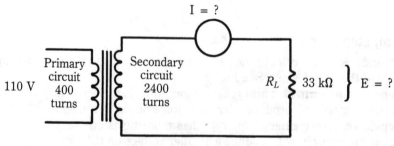

**6-2**   Step-up transformer.

# Transformers in impedance matching

When an impedance mismatch exists, power losses occur. When impedance is properly matched, maximum power transfer is possible. Transformers are often used to match the impedance. The relationships between impedance and turns ratio is:

$$\frac{Z_p}{Z_s} = \left(\frac{N_p}{N_s}\right)$$

where

$N$ = number of turns
$Z$ = impedance
$p$ = primary
$s$ = secondary

**Example**  The output resistance of an audio amplifier is 4000 $\Omega$. What is the primary-to-secondary turns ratio of a transformer designed to match the amplifier to a 10-$\Omega$ speaker?

$$\frac{4000}{10} = 20.1 \text{ (turns ratio)}$$

# Transformer power losses

Although transformers are used to promote maximum power transfer, there are some inherent power losses in transformers:

- Eddy currents  Eddy currents are iron core losses in a transformer. They cause heat dissipation and therefore power loss in the iron core. A laminated iron core reduces eddy currents by increasing the resistance in the core.
- Copper losses  These are power losses caused by the resistance of the windings. Larger diameter wire will reduce the resistance and therefore reduce these losses.
- Hysteresis Loss  These losses result from the changing direction of flux in the iron core of the transformer. It takes energy to reverse the direction of the magnetic field.

# Study questions
# Transformers

1. A _____ transformer has more secondary than primary windings.

2. What is the relationship between voltage, current, and number of turns in the primary and secondary of a transformer?

3. What is the induced voltage in the secondary of a transformer where:
   a. There are 200 turns on the primary
   b. There are 1200 turns on the secondary
   c. The voltage applied to the primary is 110 V

4. What is the current flow through a 50-Ω resistor that is connected across the secondary of the transformer in the problem above?

5. What is the current flow through a 200 kΩ resistor connected across the secondary of the above transformer?

6. The output resistance of an audio amplifier is 7500 Ω. What turns ratio is required to match this output to a 12-Ω speaker?

7. _____ are iron core losses in a transformer that cause heat and loss of power.

8. _____ losses are power losses caused by the electrical resistance of the transformer windings.

# 7

# Semiconductors

Pure germanium does not have free valence electrons available. The four valence electrons are all tightly bonded together. This makes germanium a good insulator. A *semiconductor* is formed when a small amount of impurity is added to a material like germanium. One such impurity is phosphorous, containing five valence electrons. When phosphorus combines with the germanium, four of the five electrons form bonds with the adjacent germanium atoms. See Fig. 7-1. This leaves one free electron per phosphorous atom. Because there are millions of free electrons in the semiconductor, electrical current can now flow through the material. This type of semiconductor is called an *n-type* because it has an excess of negative electrons. If the impurity contains only three outer valence electrons, the result will be a deficiency of electrons in the atomic structure. These appear as positive *holes*. Current flows in these materials by an apparent movement of the holes. This is called *p-type* material.

## The semiconductor diode

Current flow takes place in both types of material (p-type and n-type). If the polarity of the battery is reversed, the direction of the current also reverses. It is when the two materials are joined together, forming a *junction*, that new properties develop. A pn junction will conduct current only in one direction. When it is *forward biased*, it acts like a short circuit (closed switch) and conducts. When it is *reverse biased*, it acts like an open circuit (open switch) and does not conduct current. About 0.7 V is required to forward bias a silicon diode, and about 0.3 V to forward bias a germanium diode. Figure 7-2 illustrates forward bias on a diode. Notice that the negative side of the battery is connected to the negative side of the diode, and that the positive side of the battery is connected to the positive side of the diode (positive-to-positive and negative-to-negative arrangement). Figure 7-3

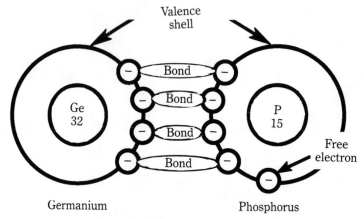

**7-1** Semiconductors.

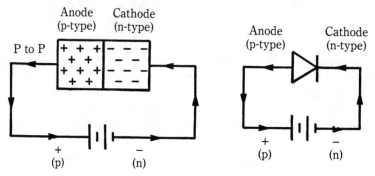

**7-2** Forward bias on diode.

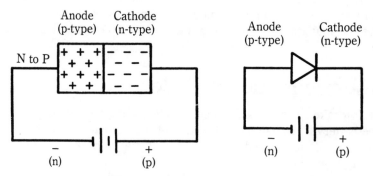

**7-3** Reverse bias on diode.

illustrates a diode that is reverse biased. Notice that the positive side of the battery is connected to the negative side of the diode and that the negative side of the battery is connected to the positive side of the diode (positive-to-negative and negative-to-positive arrangement). Current only flows when the diode is forward biased.

# The transistor

The diode has two materials, creating one junction. The transistor contains three materials, forming two junctions. Because of the two types of semiconductor material and the junctions they form, the name bipolar junction transistor (*BJT*) is often used. The two basic types of junction transistors are the pnp and the npn transistor, depending on the arrangement of semiconductor materials. Figure 7-4 shows the circuit symbol for these two basic transistors. Figure 7-5 shows how the transistor resembles triode vacuum tube as follows:

- Emitter (E)   The emitter is similar to the cathode of a tube, in that it emits the electrons.
- Base (B)   The base is similar to the control grid of a tube, in that it controls the flow of electrons.
- Collector (C)   The collector is similar to the plate of a tube in that it collects the electrons.

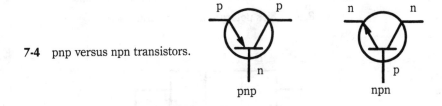

**7-4**   pnp versus npn transistors.

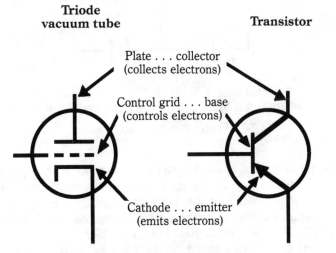

**7-5**   Vacuum tube versus transistor.

**Transistor bias**   The bias voltage applied to the base of the transistor determines how the transistor will operate. Transistors are generally forward biased on the emitter-base junction, and reverse biased on the collector-base junction. The bias polarity, as indicated by batteries in diagrams in this book, is very important. A question you might encounter on an FCC test could be to select the transistor diagram that has proper battery placement. Figure 7-6 illustrates proper bias polarity on an npn transistor. Figure 7-7 shows proper bias polarity on a pnp transistor.

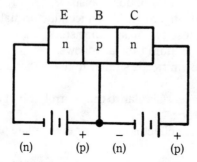

(Forward bias)                          (Reverse bias)
(n of battery to n of transistor)   (p of battery to n of transistor)

**7-6** Proper bias polarity on an npn transistor.

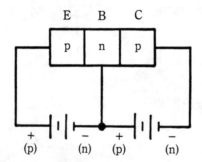

(Forward bias)                          (Reverse bias)
(p of battery to p of transistor)   (n of battery to p of transistor)

**7-7** Proper bias on a pnp transistor.

The amount of voltage applied to the two junctions is an important consideration. The emitter-base junction voltage must be considerably less than the voltage applied to the collector-base junction. If they are equally biased, the transistor will not operate. Figure 7-8 illustrates proper bias level and Fig. 7-9 shows improper bias level. In Fig. 7-9, the emitter-base bias is too high.

The amount of voltage applied to the base of the transmitter determines whether it will function as an amplifier or a transistor switch. The transistor requires at least 0.7 V to activate it. This is called the cutoff voltage, because any voltage less than this will cut off the amplification in the transistor. Below cutoff, the transistor acts like an open switch or a class C amplifier. When the transistor is biased at 0.7 V, it acts as a class B amplifier (various amplifier classes are discussed in Chapter 10).

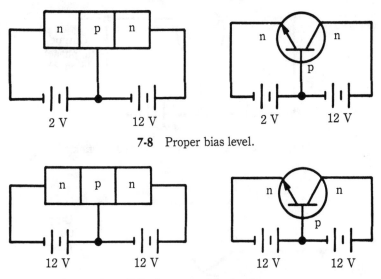

**7-8**  Proper bias level.

**7-9**  Improper bias level.

# Transistor circuit configurations

There are three basic arrangements of transistor circuits. Each has its own set of characteristics of input and output impedance ($Z$). Of the three circuits, only the *common-emitter* arrangement produces a phase reversal. That is, the output signal is 180° out of phase with the input signal. Table 7-1 summarizes the circuit configurations and their characteristics.

**Table 7-1. Characteristics of transistor circuit configurations.**

| Circuit | Input $Z$ | Output $Z$ | Phase reversal |
|---|---|---|---|
| Common base | low | high | no |
| Common collector | high | low | no |
| Common emitter | low | medium | yes |

Figure 7-10 illustrates the proper biasing on a common-base amplifier. This configuration offers a low input impedance but a high output impedance. The output is in phase with the input. Figure 7-11 shows proper biasing on a common-emitter amplifier. This amplifier offers a low input impedance and a medium-high output impedance. The output is inverted (180° phase shift). Figure 7-12 illustrates proper biasing of a common-collector amplifier. This configuration offers a high input impedance and low output impedance. No phase reversal occurs as the signal is amplified. Please note that pnp transistors are used in these illustrations. If npn transistors were used, all battery polarities would need to be reversed.

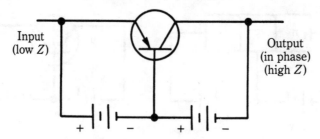

**7-10**   Proper bias polarities on a common-base amplifier.

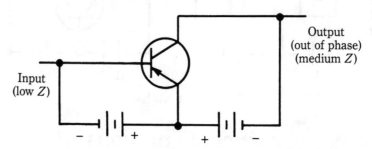

**7-11**   Proper bias polarities on a common-emitter amplifier.

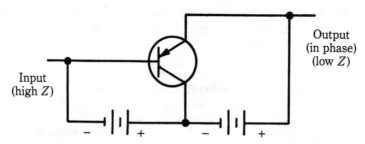

**7-12**   Proper bias polarities on a common-collector amplifier.

# Effect of positive input signal on transistor amplifiers

There are two important things to learn in this section. One is that the common emitter amplifier provides a phase reversal, but the common base and common collector amplifiers do not. The other important thing to learn is what effect a positive input signal has on various types of amplifiers. Don't get bogged down in the detail, but establish a firm understanding quickly. Note that Figs. 7-13 through 7-21 are unique to this book and will probably not be seen on an FCC test. However, firmly commit all other figures in this chapter to memory.

When a signal is applied to the input of a transistor amplifier, the bias voltage is either increased or decreased, depending upon the polarity of the input signal, the amplifier configuration, and the type of transistor.

The input will either add or subtract from the forward bias applied to the emitter-base junction. A positive input signal would add to the forward bias if that bias is positive. However, a positive input signal will subtract from the bias if the bias is negative. For example, consider what happens when a positive signal is applied to the pnp common-base amplifier (Fig. 7-10). The positive input signal will add to, and therefore increase, the positive forward bias on the transistor. This will, in turn, increase the collector current. When a positive signal is applied to the common-emitter amplifier (Fig. 7-11), the input signal opposes the forward bias and reduces it. This reduces the collector current. A positive signal applied to the common-collector (Fig. 7-12) will also oppose the polarity of the bias, reducing the forward bias and the collector current. Table 7-2 summarizes the effect of a positive input signal on the npn, and Table 7-3 summarizes the effect of a positive input signal on the pnp transistor. The three basic circuit configurations are considered.

**Table 7-2. The effect of a positive input signal on an npn transistor.**

| Circuit configuration | Forward bias | Collector current |
|---|---|---|
| Common base | decreases | decreases |
| Common emitter | increases | increases |
| Common collector | increases | increases |

**Table 7-3. The effect of a positive input signal on a pnp transistor.**

| Circuit configuration | Forward bias | Collector current |
|---|---|---|
| Common base | increases | increases |
| Common emitter | decreases | decreases |
| Common collector | decreases | decreases |

Once the effect of various input signals is understood, you can understand why the common-emitter arrangement causes a phase reversal, but the common-base and common-collector arrangements do not. Think of the transistor as a variable resistor. The forward bias determines its resistance setting. As the forward bias on the transistor is increased, less resistance in the transistor allows increased collector current to flow through the load resistor. When the forward bias is reduced, the resistance increases, restricting collector current flow. For the next three examples, assume that the load resistance is 1 kΩ. RQ represents the transistor, acting as a variable resistor.

In these examples, you will determine what happens when a positive-going signal is applied to the input of the three basic amplifier configurations. Figure 7-13 illustrates a common-base amplifier at the onset of a positive input signal. $R_Q$ (transistor) is 2 kΩ. The series circuit that is formed with $R_L$ and $R$, totals 3 kΩ. Current $I = E/R$ is 12 V/3000 Ω, or 0.004 A. The voltage drop across RL can then be calculated by $E = I \times R$, or $0.004 \times 1000 \ \Omega = 4$ V. This voltage drop across RL reduces the measured voltage, as measured from the output terminal to ground, from $-12$ V to $-8$ V. As the input signal moves more positive, as shown in Fig. 7-14 the forward bias is reinforced, or added to. This causes the transistor (RQ) to offer less resistance, allowing more current flow through the collector circuit and RL. With more current flowing through RL, more voltage is dropped. In this case, use Ohm's law to determine that 6 V is dropped across RL. This reduces the $-12$ V to $-6$ V. Note that this change is in the positive direction (less negative). If the positive input signal becomes sufficient to reduce RQ to zero ohms, maximum current will flow in the collector circuit (Fig. 7-15). In this case, all the voltage is dropped across RL, leaving zero volts at the output. This is also a move in the positive direction. So, it can be seen that as the input signal moves in a positive direction, so does the output signal. In other words, the output is in phase with the input. Generally speaking, when the forward bias is reinforced by an input signal of like polarity, all currents in the transistor increase and more voltage is dropped across RL. Pay close attention to the battery polarity in the transistor output circuit.

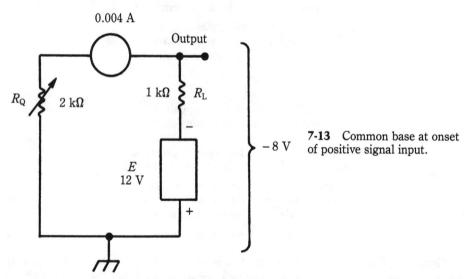

**7-13**   Common base at onset of positive signal input.

Now look at the common-emitter configuration. Although the polarity in the output circuit is the same as in the common base, the polarity in the input circuit is opposite. This means that as a positive signal is applied to the input, the bias is counteracted. The pnp transistor in the common-emitter arrangement requires a negative bias on the base (Fig. 7-11). Assume that Fig. 7-16 represents the amplifier at the onset of a positive-going signal. Notice that zero volts is present from the

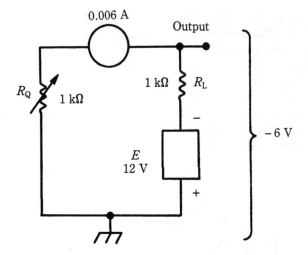

**7-14** Common base with mid-range, positive-going input signal.

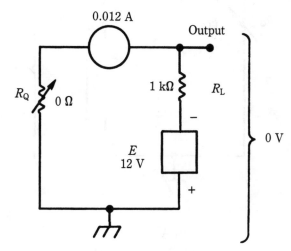

**7-15** Common base with full-range, positive-going input signal.

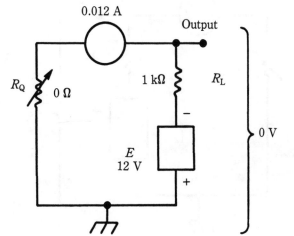

**7-16** Common emitter at onset of positive-going input signal.

output terminal to ground. As the input signal moves in a positive direction, the forward bias is reduced, increasing the resistance of the transistor (RQ). This reduces all currents and the voltage drop across RL, making the output signal – 6 V (Fig. 7-17). When the positive input signal is sufficient to increase the value of RQ value to 2 kΩ, even less voltage is dropped across RL as shown in Fig. 7-18. The output becomes – 8 V, a move in the negative direction. So, when the input signal is positive-going, the output is negative going. This means that there is a 180° phase shift, making the input and output out of phase. This only occurs in the common-emitter configuration.

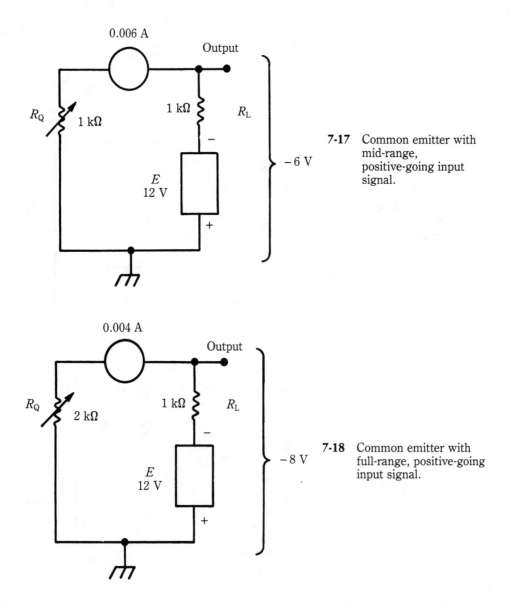

**7-17**   Common emitter with mid-range, positive-going input signal.

**7-18**   Common emitter with full-range, positive-going input signal.

The final configuration is the common-collector (Fig. 7-12). Notice that the input polarity is the same as the common emitter, but the output circuit polarity is opposite. Figures 7-19, 7-20, 7-21 illustrate how the output signal moves positive as the input signal moves positive (no phase shift).

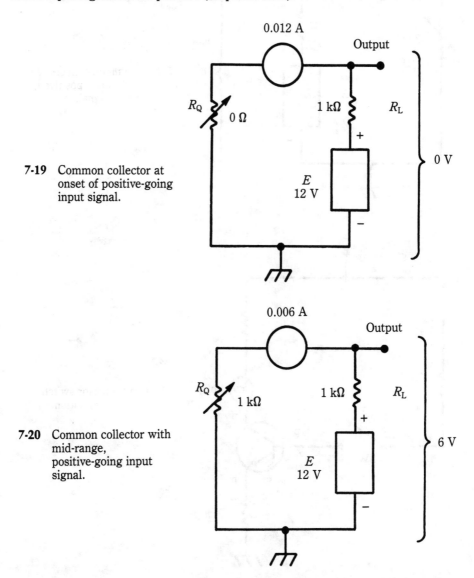

**7-19** Common collector at onset of positive-going input signal.

**7-20** Common collector with mid-range, positive-going input signal.

# The transistor as a switch

Figure 7-22 illustrates an npn transistor designed to operate as a switch. When the mechanical switch (SW) is open, no bias is applied to the base of the transistor (Q1). To forward bias Q1, a positive potential of at least 0.7 V (cutoff voltage) must

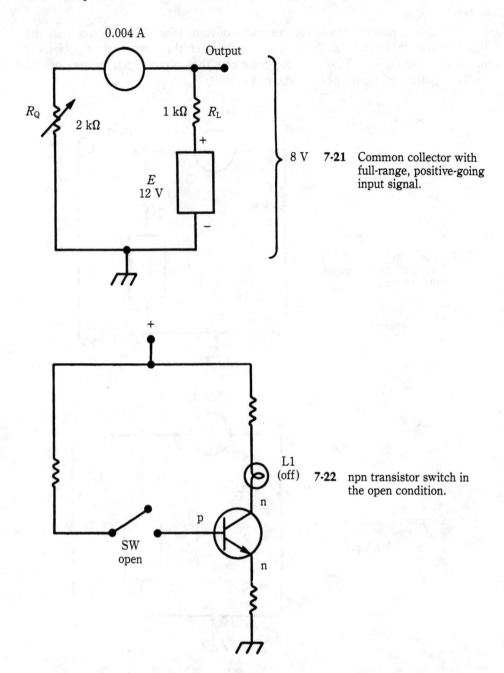

0.004 A

Output

$R_Q$

2 kΩ

1 kΩ   $R_L$

+

$E$
12 V

−

8 V   **7-21**   Common collector with
full-range, positive-going
input signal.

+

L1
(off)   **7-22**   npn transistor switch in
the open condition.

n

p

n

SW
open

be applied to its base. Hence, Q1 is below cutoff, acting like an open switch, or
open circuit for the L1 (lamp) circuit. In this case, L1 is off. When SW is closed, as
in Fig. 7-23, positive bias is applied to the base of Q1. If that bias is sufficiently
high, Q1 will go into *saturation*, meaning that its resistance becomes minimal. A

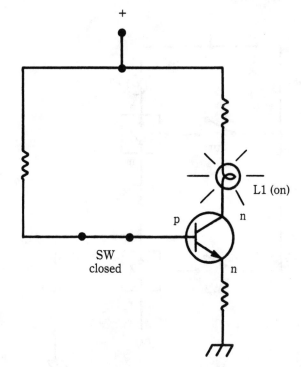

**7-23**  npn transistor switch in the closed condition.

saturated transistor acts like a closed switch (short circuit). Assuming Q1 became saturated, current flows from ground through Q1, through L1 and RL to $V_{cc}$, turning the lamp (L1) on. To further illustrate this, Figs. 7-24, 7-25, and 7-26 show how two transistor switches can interact in a circuit. SW1 and SW2 are mechanical switches, and Q1 and Q2 are transistor switches. The npn transistor is substituted with the symbol of a mechanical switch for illustration purposes. When a mechanical switch is closed, the transistor is forward biased into saturation, causing it to act like a closed switch. Figure 7-24 shows that with SW1 and SW2 both open, no current flows in the circuit and both lamps are off. Figure 7-25 shows that when SW1 is closed, Q1 is on, allowing a path for current to flow through both lamps. Figure 7-26 shows what happens when SW2 is also closed. In this case, both transistor switches are on, providing a low resistance path for current flow. L2 is bypassed, leaving only L1 on.

# Operating point

The transistor bias determines its operating point. As discussed, at one end of the range is cutoff (0.7 V), below this the transistor does not operate. A transistor at cutoff acts like an open switch. At the opposite end of the operating range is saturation. A saturated transistor acts like a closed switch. Between these two extremes lies the transistor amplification range.

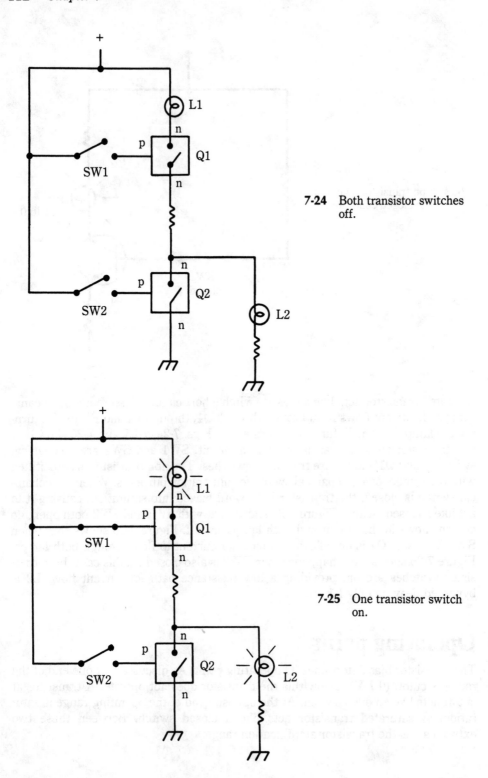

**7-24** Both transistor switches off.

**7-25** One transistor switch on.

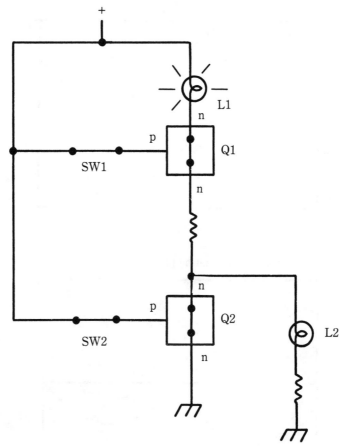

**7-26**  Both transistor switches on.

To understand how the bias effects the operating range, refer to Figs. 7-27, 7-28, and 7-29. Imagine this to be a half-full water tank. It is the water level that determines the operating Point of the transistor. The water level is the bias. If the edge of the tank is bumped, a wave will travel across to the other side of the tank. Because the water level is midrange, the wave will travel without risking striking either the top or the bottom of the tank. This is analogous to a class A amplifier. If the bias is set too high, the operating point will be close to saturation. As shown in Fig. 7-27, the top of the wave is clipped.

Figure 7-29 shows what happens when the operating point is set too low. The lower portion of the wave is clipped. Class B amplifiers are biased at cutoff (0.7 V). In this case, only one half of the input sine wave is amplified. The transistor is cut-off for the other one-half cycle. Class C amplifiers are biased even lower (zero or less). In such an amplifier, only a small portion of the signal peaks is amplified. The characteristics of various classes of amplifiers is covered in more detail in Chapter 10.

Operating point

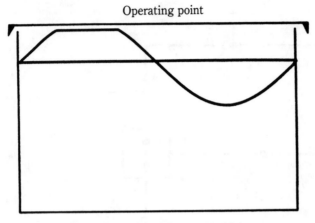

**7-27**   Operating point set too high.

Operating point

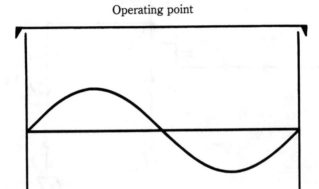

**7-28**   Operating point set mid-range.

Operating point

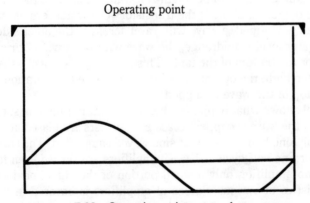

**7-29**   Operating point set too low.

### Transistor damage

Transistors are very sensitive to heat. The increased temperature causes the *collector current* to increase. This will eventually damage the transistor if allowed to continue.

## The field-effect transistor

The FET (field-effect transistor) is a transistor that is characterized by a very high input impedance as opposed to a low input impedance in a pnp or npn bipolar junction transistor. The FET also has a low noise figure, making it excellent for use in preamplifier circuits. Its high dynamic range enables it to be used up to the ultrahigh frequency range.

Figure 7-30 illustrates the electronic symbol for the FET. This type is more specifically called an n-channel junction FET or JFET. Figure 7-31 illustrates a metal oxide semiconductor field-effect transistor (MOSFET). The MOSFET is particularly sensitive to static electricity. It can be permanently damaged from static electricity from your fingers. Use care when handling them.

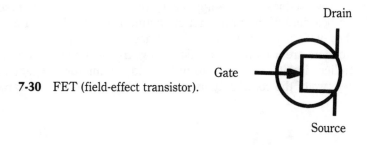

**7-30**   FET (field-effect transistor).

**7-31**   MOSFET (metal-oxide semiconductor field-effect transistor).

## The zener diode

The zener diode resembles its vacuum tube counterpart—the gaseous tube rectifier. The zener diode is more reliable, less expensive, and therefore more commonly used. The zener diode functions in the reverse-bias mode. As the reverse bias increases, a breakdown voltage is eventually reached. When this point is reached, the diode conducts and produces a stable reference voltage. The two primary uses are:

- Voltage regulator   When used as a voltage regulator, the zener diode provides a nearly constant dc voltage output in the presence of varied load resistance changes or changes in input voltage.
- Voltage-reference diode   In this capacity, the zener diode produces a voltage drop across its junction when a certain current flows through it in the proper direction. Figure 7-32 illustrates the electronic symbol for the zener diode.

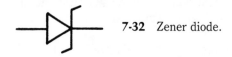

**7-32**   Zener diode.

# The varactor

The *varactor* (voltage-variable capacitor) is a diode constructed with special impurities that increases the capacitance at its junction. The capacitance can be controlled by the reverse-bias voltage applied to the diode. As the reverse bias is increased, the boarders of the pnp junction, acting like two plates of a capacitor, widen. As they widen, the capacitance decreases, just as it does in any other capacitor. Varactors are used in electronic tuning, frequency modulation, and frequency-multiplier circuits, and operate well up to the microwave region. Figure 7-33 illustrates two of the commonly used electronic symbols for the varactor.

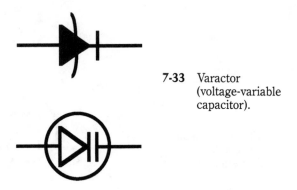

**7-33**   Varactor (voltage-variable capacitor).

# Operational amplifiers

The op-amp (operational amplifier) is an amplifier whose characteristics are determined by components external to the amplifier. Several op-amps can be built on a small integrated circuit, or *chip*. The schematic symbol for the op-amp consists of a triangular shape with two inputs and one output. The ( – ) input is called the *inverting input*. When the input is at the ( – ) input, it is called an inverting op-amp, as shown in Fig. 7-34. If a positive voltage is applied to the inverting input, a negative voltage will appear at the output. If a sine wave signal is applied to the inverting input, the output sine wave will be 180° out of phase with the input. If the input is at the (+) input, it is called a noninverting op-amp, as illustrated in Fig. 7-35. If a positive voltage is applied to the noninverting input (+), a positive voltage will appear at the output. A sine wave applied to the noninverting input will be in phase at the output.

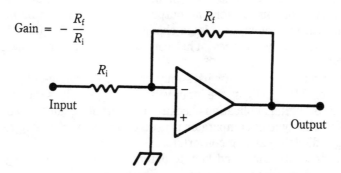

$$\text{Gain} = -\frac{R_f}{R_i}$$

**7-34** Inverting op-amp (operational amplifier).

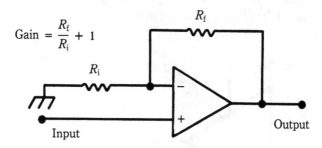

$$\text{Gain} = \frac{R_f}{R_i} + 1$$

**7-35** Noninverting op-amp.

Op-amps are very high gain devices. External resistance are used to lower the gain. The gain of an inverting op-amp is calculated as follows:

$$\text{Gain} = -\frac{R_f}{R_i}$$

where
    $R_f$ = feedback resistance
    $R_i$ = input resistance

Please note that the minus sign denotes that it is an inverting op-amp, resulting in a 180° phase shift.

The gain of a noninverting op-amp is calculated as follows:

$$\text{Gain} = \frac{R_f}{R_i} + 1$$

Take note that in both types of op-amps, the feedback resistor is connected to the inverting input. Remember this in case the polarity signs are not shown on a diagram.

# Comparators

Op-amps can be used as comparators. A comparator compares two voltages. One voltage is a fixed reference voltage ($V_{ref}$), and the other is a variable input voltage ($V_{in}$). In one type of comparator, the output swings from maximum negative to maximum positive as the input voltage changes. This type requires both positive and negative dc supply voltages. This can be costly and in some applications unnecessary.

Another type of comparator uses only one supply voltage ($V_{cc}$). The output swings between 0 V and $V_{cc}$ as the input voltage changes. This type of comparator is well suited for digital applications because the output equates to 0 or 1. Because of this quality, this type of comparator is used extensively in analog-to-digital converters and in digital-to-analog converters.

You will need to understand how to determine the output voltage, when the input and reference voltages are given, as illustrated in Fig. 7-36. The reference voltage can be positive or negative. To further complicate the matter, the reference voltage can enter in at either the inverting ( – ) or the noninverting ( + ) input of the comparator. This can be simplified in the following manner.

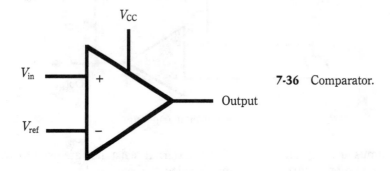

7-36  Comparator.

Do not be concerned which is the input voltage and which is the reference voltage. All you need to look at is how much voltage is present at the ( – ) input and the ( + ) input. In all cases, if the voltage entering at the ( – ) input is more positive than the voltage entering at the ( + ) input, the output will be 0 V. If the voltage entering at the ( + ) input is more positive than the voltage present at the ( – ) input, the output will be equal to $V_{cc}$. Notice, in Table 7-4, that when the voltage at ( – ) is – 1 V and the voltage at ( + ) is 2 V, the output is equal to $V_{cc}$. When the voltage at the ( – ) input is at V and the voltage at the ( + ) input is – 2 V, the output is 0. Finally, with – 1 V at the ( + ) input and – 3 V at the ( – ) input, the output is equal to $V_{cc}$.

If you have such a problem on your FCC test, simply draw a vertical line on your paper. Then number from 0 up to + 3 V and from 0 down to – 3 V. Place a +

Table 7-4. Comparator output.

| | | Output | | |
|---|---|---|---|---|
| | | $V_{cc}$ | 0 V | $V_{cc}$ |
| | +4 | | | |
| V | +3 | | | |
| O | +2 | + input | | |
| L | +1 | | – input | |
| T | 0 | | | |
| A | –1 | – input | | + input |
| G | –2 | | + input | |
| E | –3 | | | – input |
| | –4 | | | |

When the voltage applied to the + input is higher, or more positive, the output is equal to $V_{cc}$. When the voltage applied to the – input is more positive, the output is 0 V.

and a – at the appropriate points on the graph that indicate the amount of voltage present at those inputs. If the + is higher, the output is equal to $V_{cc}$. If the – is higher, the output will be 0 V.

# Study questions
# Semiconductors

1. A pn junction has a _____ resistance and conducts when it is _____ biased.

2. A pn junction has a _____ resistance and does not conduct when it is _____ biased.

3. Forward bias is where the positive side of the battery is connected to the _____ type side (anode) and the negative side of the battery is connected to the _____ type side (cathode) of the diode.

4. The emitter of a transistor is similar to the _____ of a vacuum tube.

5. The base of a transistor is similar to the _____ of a vacuum tube.

6 The collector is similar to the _____ of a vacuum tube.

7. In a transistor, the emitter-base junction is _____ biased.

8. The collector-base junction is _____ biased.

9. The emitter-base junction bias must be _____ than the collector-base bias.

10. The common-base configuration has a _____ input impedance and a _____ output impedance.

11. Of the three basic transistor configurations, which one produces a phase reversal of the signal? _____.

12. What happens to the current flow when a positive pulse is applied to a pnp transistor in a common-base arrangement?_____.

13. Summarize what happens when a positive signal is applied to various transistor circuits.

14. The field-effect transistor (FET) is characterized by a _____ input impedance.

15. One use for the zener diode is _____.

16. The _____ is a specialized diode. Its internal capacitance can be controlled by the voltage applied.

17. An op-amp is an amplifier whose characteristics are determined by components _____ to the amplifier.

18. How is the gain of an inverting op-amp calculated?

19. What is the output of the comparator in Fig. 7-36, where the reference voltage is $-1$ V, the input voltage is 2 V, and the $V_{cc}$ voltage is 5 V?

# 8
# Power supplies

The transformer is an effective source of alternating current. A device that converts ac to dc is the diode. When a semiconductor diode is forward biased, it offers a low electrical resistance and conducts. When it is reverse biased, it offers a high resistance and does not conduct.

Figure 8-1 illustrates a forward-biased diode. Notice that current flows in this circuit. Figure 8-2 illustrates a reverse-biased diode. Current does not flow in this circuit.

Notice diodes D1 and D2 in Fig. 8-3. By carefully examining the battery polarity and diode positioning, you can see that D1 is forward biased and D2 is reverse biased. This means that current flows through R1, D1 and R2. D2 acts like an open switch, removing R3 from the circuit.

The capability of allowing electrical current to flow in one direction is called *rectification*. Rectifiers are arranged in various configurations that allow part or all of the ac wave to be converted into dc.

## The half-wave rectifier

If the secondary of a transformer is connected to a single diode, as shown in Fig. 8-4, only half of the ac wave will be rectified. It can be seen that as the secondary changes polarity, the diode is forward biased for half of the cycle and reverse biased for the other half of the cycle. This results in only the positive half conducting through the diode. A pulsating dc is formed at the load resistor. The pulsating dc is then sent to a filter circuit to remove the amplitude variations (ripple).

Figure 8-4 shows the single-diode, half-wave rectifier circuit. Also, in that illustration, note the full sine wave input. The output waveform is composed only of the positive half of the sine wave. The diode is reverse biased for the negative portion of the sine wave.

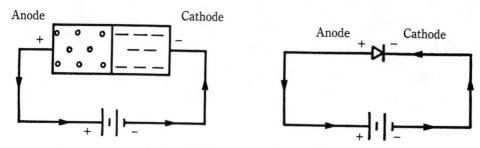

**8-1** Forward-biased power supply diode.

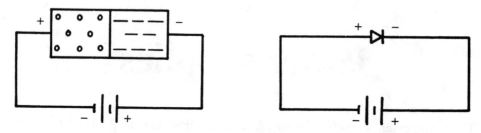

**8-2** Reverse-biased power supply diode (No current flows).

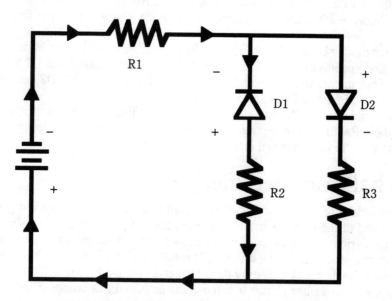

**8-3** Diode conduction (D2 acts like an open switch).

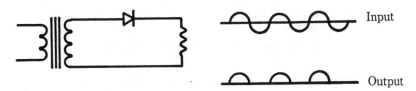

**8-4** Half-wave diode rectifier with input and output waveforms.

# The full-wave rectifier

If a second diode is added to the secondary of the transformer, as shown in Fig. 8-5, both half cycles are converted into dc. The pulsating dc output of a full-wave rectifier is easier to filter because the pulsation rate is twice as high as the rate of the half-wave rectifier. Higher frequencies are easier to filter.

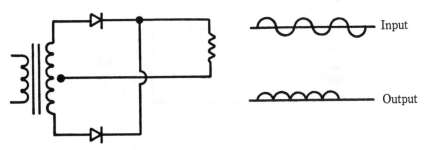

**8-5** Full-wave rectifier with input and output waveforms.

### Protection of diodes

Semiconductor diodes have a lower PIV (peak inverse voltage) rating than vacuum tube diodes. When the peak inverse voltage rating is higher than the maximum rating for a single diode, they can be connected in series. The voltage should be distributed equally across all the diodes in series to avoid damage. This is accomplished by *shunt resistors*, also called *equalizing resistors*. The shunt resistors protect the diodes from excessive PIV by distributing it equally, thus maintaining a constant voltage across each diode. The diodes can be damaged by high-voltage transients. *Shunt capacitors* are used to absorb and equalize the voltage transients. The capacitors have a low reactance to high frequency transients, and therefore provide a low resistance path from one diode to another. Shunt capacitors act like shock absorbers. Figure 8-6 illustrates these two protective measures.

### Power supply regulation

A power supply with good *regulation* will maintain a constant output in the presence of varying load conditions. One method of regulating a power supply is by

Shunt resistors are used to protect the sensitive
diodes from excessive PIV (peak inverse voltage).

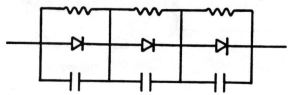

Shunt capacitors protect diodes from voltage transients.

**8-6** Diode protection.

using a zener diode. The zener diode maintains a constant output across it, even
when the load conditions change. Zener diodes are rated from 4 to 200 V. If a 9-V
zener is used a regulated 9 V supply would be formed. Figure 8-7 illustrates a
zener diode used to produce a constant 9-V power supply output.

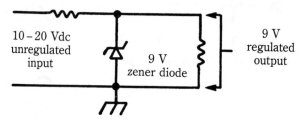

**8-7** Simple zener diode regulator circuit.

Figure 8-8 shows a series voltage regulator using a zener diode in the base of
the transmitter. The unregulated output of a rectifier circuit (not shown) is connected to a capacitor input filter. The filtered output is applied to the series regulator. The zener diode fixes the base voltage at a constant level. For example, if a 6-V
zener is used, the transistor base is supplied with a constant 6 V. To determine how
much voltage appears across the load resistor (RL), simply subtract 0.7 V from the

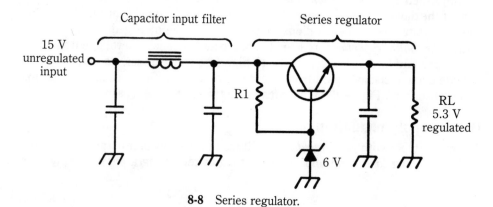

**8-8** Series regulator.

rated voltage of the zener diode. This is because of the 0.7 V breakdown voltage between the base and the emitter of the transistor. Therefore, a 12-V zener produces a load voltage of $12 - 0.7 = 11.3$ V. A 6-V zener produces a load voltage of $6 - 0.7 = 5.3$ V. This can be rounded off to roughly 5 V. Note that if resistor R1 is open, or not present, the circuit will not operate.

# Voltage dividers

Electronic circuits often require several different voltages. A single power supply used in conjunction with a voltage divider can produce several voltage sources. The voltage divider consists of a series of resistors that allow individual voltage drops. As the voltage is dropped across each resistor, the power supply voltage is literally divided. The following circuit is a voltage divider with one source voltage (Fig. 8-9).

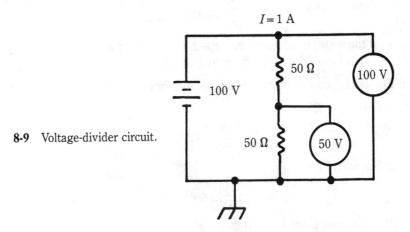

**8-9** Voltage-divider circuit.

Notice that a simple application of Ohm's law is used to calculate the voltage drops. However, when two voltage sources are present, they both influence the circuit independently. For example, in the following circuit, calculate the voltage at point P with respect to ground (Fig. 8-9). Figure 8-10 shows an equivalent circuit.

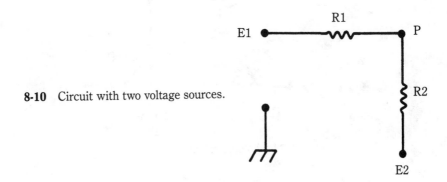

**8-10** Circuit with two voltage sources.

# The superposition theorem

Calculating the voltage at certain points in a circuit, where one battery source is used, is straightforward. In some circuits, two voltage sources can exert an influence at various points. The superposition theorem states that the voltage at a point is equal to the algebraic sum of the voltages produced by each source acting independently at that point.

In a circuit like Fig. 8-10, you can calculate the voltage at point P as follows:

1. Calculate the influence of $E_2$ as follows:

$$\frac{R_1}{R_{total}} \times E_2 = \underline{\hspace{2cm}} \text{ volts}$$
$$R_{total} = R_1 + R_2$$

2. Calculate the influence of $E_1$ as follows:

$$\frac{R_2}{R_{total}} \times E_1 = \underline{\hspace{2cm}} \text{ volts}$$

3. Add the voltage values from steps 1 and 2. Their algebraic sum is the voltage as measured from point P to ground.

In Fig. 8-11, you can calculate the voltage from P to ground as follows:

1. Influence of E2:

$$\frac{70,000}{100,000} \times -30 \text{ V} = -21 \text{ V}$$

2. Influence of E1:

$$\frac{30,000}{100,000} \times 40 \text{ V} = 12 \text{ V}$$

3. Voltage from point P to ground:

$$(-21 \text{ V}) + (12 \text{ V}) = -9 \text{ V}$$

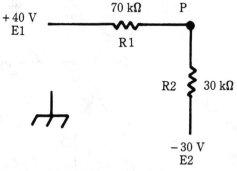

**8-11** Voltage calculation.

# Study questions
# Power supplies

1. A full-wave rectifier converts the full sine wave into dc. It therefore has a ripple frequency of _____ that of a half-wave rectifier.

2. Semiconductors diodes have a _____ PIV rating than vacuum tubes.

3. Shunt resistors protect series diodes from _____.

4. Shunt capacitors protect series diodes from _____.

5. A 12-V zener diode will regulate the output of the power supply to a constant _____ volts.

6. Good power supply regulation means the output voltage remains constant in the presence of _____.

7. In the following circuit, what is the total current? (Fig. 8-12).

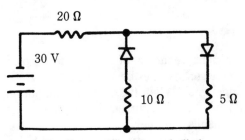

**8-12** Current flow through diodes.

# 9

# AM and FM receivers

Figure 9-1 shows a block diagram of a single-conversion AM superheterodyne receiver. The text associated with that diagram gives a brief overview of the receiver. It is followed by a more detailed description of the individual components.

## AM receivers

**Antenna**  The resonant frequency of the antenna is primarily determined by its physical length. A receiver operates best if the antenna is matched with the frequency of the receiving signal. When the incoming signal reaches the antenna, a minute current is induced. This RF current is coupled to the RF amplifier of the receiver through the transmission line.

**Radio-frequency amplifier**  In a transmitter, class C RF amplifiers are used because the *flywheel effect* of the tank circuit reproduces the original waveform. However, in a receiver, class A amplification is necessary because of its high fidelity. Whenever an RF signal containing audio intelligence is to be amplified, high fidelity is necessary. The amplified RF signal is coupled to the mixer where it will be mixed with the local oscillator signal.

**Local oscillator**  The local oscillator is a radio-frequency oscillator that generates a signal at 455 kHz higher than the incoming signal. Its frequency-controlling capacitor is mechanically coupled to the frequency-controlling capacitor of the RF amplifier so that the signal generated is always 455 kHz above the frequency of the incoming signal. The local oscillator signal is sent to the mixer.

**Mixer**  The mixer is also called the first detector or the converter. This is where the conversion of frequency occurs in the single-conversion receiver. The two input signals to the mixer are the local oscillator and the incoming signal from the RF amplifier. When two signals are mixed or *heterodyned*, two new signals are

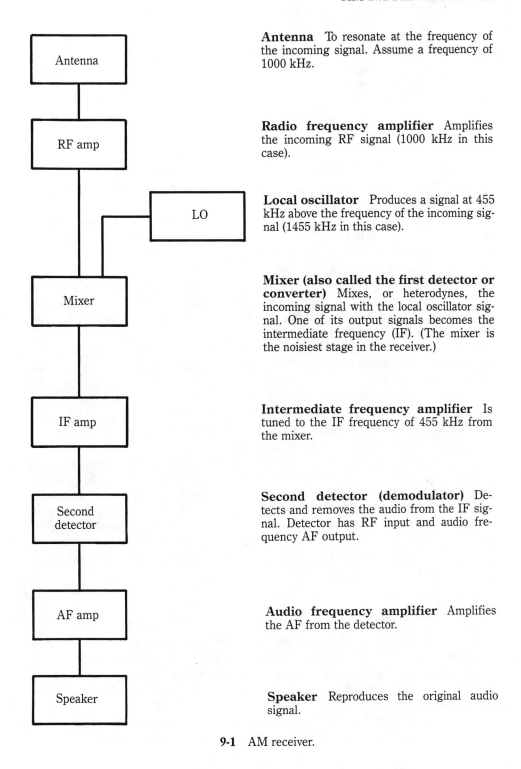

**Antenna** To resonate at the frequency of the incoming signal. Assume a frequency of 1000 kHz.

**Radio frequency amplifier** Amplifies the incoming RF signal (1000 kHz in this case).

**Local oscillator** Produces a signal at 455 kHz above the frequency of the incoming signal (1455 kHz in this case).

**Mixer (also called the first detector or converter)** Mixes, or heterodynes, the incoming signal with the local oscillator signal. One of its output signals becomes the intermediate frequency (IF). (The mixer is the noisiest stage in the receiver.)

**Intermediate frequency amplifier** Is tuned to the IF frequency of 455 kHz from the mixer.

**Second detector (demodulator)** Detects and removes the audio from the IF signal. Detector has RF input and audio frequency AF output.

**Audio frequency amplifier** Amplifies the AF from the detector.

**Speaker** Reproduces the original audio signal.

**9-1** AM receiver.

produced. They are the sum and the difference of the two signals that were mixed together. At that point, four signals exist: the two original signals and the two new signals. The signal of interest is the difference signal. If the incoming RF signal is 1000 kHz, the local oscillator would be 1455 kHz. When the two are mixed together the following signals are present:

> 1000 kHz incoming signal
> 1455 kHz local oscillator signal
> 2455 kHz sum of the two signals
> 455 kHz difference of the two signals

The 455 kHz difference frequency is the intermediate frequency and is coupled to the IF amplifier circuit.

**Intermediate frequency amplifier**   The 455 kHz difference frequency from the mixer is amplified in the IF amplifier. This lower frequency is easier to work with; in addition, the IF amplifier can be designed for peak performance at one frequency. This results in improved selectivity and gain.

**Second detector**   This is called the second detector because the mixer is sometimes referred to as the first detector. This is the demodulator circuit where the audio intelligence is extracted from the IF signal. The second detector has two outputs: audio output that goes to the audio amplifier stages and an AVC (automatic volume control) output that is coupled to preceding stages of the receiver.

**Automatic volume control (AVC)**   The AVC circuit maintains the receiver volume at a constant level. As the strength of the incoming signal increases, the AVC reduces the overall gain of the receiver. As the strength of the incoming signal decreases, the AVC increases the gain of the receiver. As a result, the volume of the receiver remains constant, regardless of the strength of the incoming signal. This variable amplification or gain is accomplished by a feedback voltage from the second detector stage. As the strength of the incoming signal increases, the feedback voltage (AVC voltage) at the second detector increases. This voltage is fed back to the RF and IF amplifier stages to reduce the gain. If a weaker station is tuned in, the feedback voltage (AVC voltage) is reduced, thus allowing the gain of the preceding stages to increase, resulting in the necessary amplification to maintain the volume. AVC is helpful in improving the signal-to-noise ratio.

**Audio frequency amplifiers**   The audio signal from the detector is amplified in this stage and sent to the speaker.

# Coupling between stages

Coupled circuits are two circuits in which the current flow in one circuit influences the current flow in the other circuit. The characteristics of coupled circuits depend on the degree of coupling between stages. This degree of coupling, or, coefficient of coupling, is determined by the inductance of the coupling coils and by the mutual inductance.

In a receiver, the degree of coupling affects the bandwidth, the Q, and the selectivity of the receiver. Loose coupling produces a relatively narrow bandwidth and high Q (Fig. 9-2). As the coupling is made tighter, the frequency response curve broadens and the Q lessens. Maximum broadening occurs at the *critical coupling* point (Fig. 9-3). If coupling is tightened beyond this point, two resonant peaks form in the curve (Fig. 9-4). This is called *split tuning* and can be reduced by loosening (reducing) the coupling. Table 9-1 summarizes the differences between tight coupling and loose coupling. Coupling between stages is designed to allow the desired signal to pass to the next stage, while preventing unwanted harmonics.

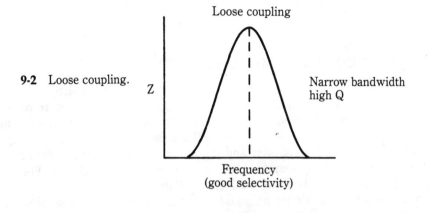

**9-2**  Loose coupling.

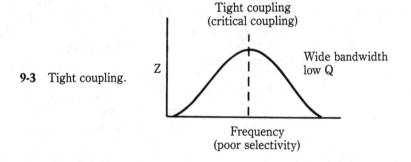

**9-3**  Tight coupling.

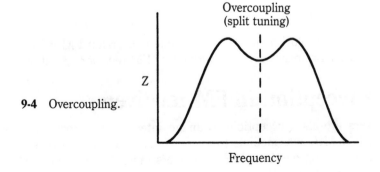

**9-4**  Overcoupling.

**Table 9-1.**
**Tight coupling versus loose coupling.**

|  | Tight coupling | Loose coupling |
|---|---|---|
| Q | low | high |
| Bandwidth | wide | narrow |
| Selectivity | low | high |
| Impedance | low | high |

# Image reception in AM receivers

Superheterodyne receivers are susceptible to interference from nearby radio transmitters that are transmitting on the image frequency of the receiver. In receivers where the local oscillator frequency is higher than the frequency of the incoming signal, the image frequency is equal to the local oscillator frequency plus the IF. For example, if the IF is 455 kHz and the incoming frequency is 1000 kHz, the local oscillator frequency is equal to the sum of the two, or 1455 kHz. The image frequency is then 455 kHz above the local oscillator, or 1910 kHz.

Notice what happens when a signal from a nearby transmitter, operating at the 1910 kHz image frequency, enters the receiver mixer. The 1910 mixes with the 1455 kHz (local oscillator signal) to produce a difference signal that is equal to the IF (455 kHz) of the receiver. Since the receiver is designed to amplify the IF signals, the unwanted signal is amplified. This interfering signal can be heard in the speaker. Figure 9-5 illustrates how the various receiver signals are related. Notice that the image frequency can be calculated by adding the IF to the LO or by adding the incoming frequency to twice the IF signal. The incoming signal used in this illustration is 1000 kHz. Please note that there is only one image frequency for each incoming frequency. The image frequency is always equal to the incoming frequency plus $2 \times$ IF. Note that some receivers have local oscillators that operate below the incoming frequency. In this case, you subtract two times the IF from the incoming frequency to calculate the image frequency.

# FM receivers

Figure 9-6 shows a block diagram of a double conversion FM superheterodyne receiver. The brief remarks give an overview of FM receiver operation.

# Image reception in FM receivers

FM receivers are also susceptible to image interference. However, in the FM receiver, a 10.7 MHz intermediate frequency is used. Figure 9-7 illustrates how a transmitter operating on 121.4 MHz (receivers image frequency) is heard on an FM radio tuned to a 100-MHz station.

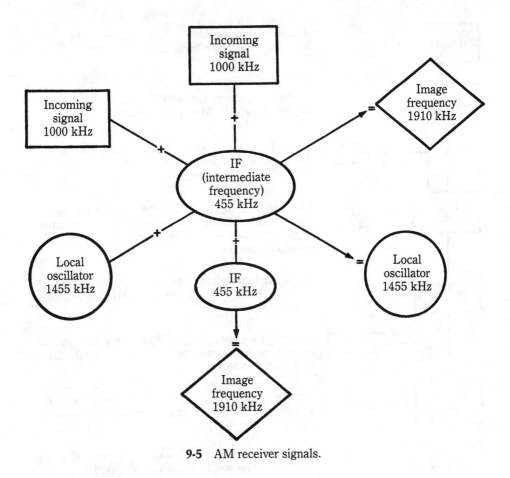

**9-5** AM receiver signals.

# Study questions
# AM and FM receivers

1. Whenever an RF signal contains intelligence, a class _____ amplifier may not be used.

2. In an AM receiver, the local oscillator is usually 455 kHz above the _____.

3. What signals are present in the mixer?
   a. _____
   b. _____
   c. _____
   d. _____

4. It is the function of the second detector to _____ the RF signal.

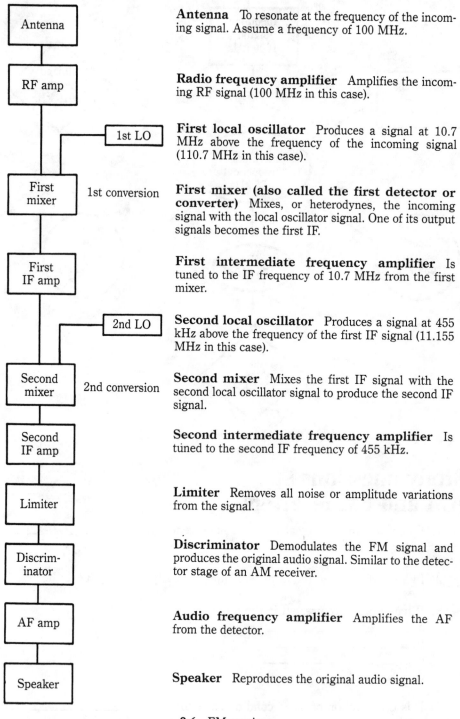

**Antenna** To resonate at the frequency of the incoming signal. Assume a frequency of 100 MHz.

**Radio frequency amplifier** Amplifies the incoming RF signal (100 MHz in this case).

**First local oscillator** Produces a signal at 10.7 MHz above the frequency of the incoming signal (110.7 MHz in this case).

**First mixer (also called the first detector or converter)** Mixes, or heterodynes, the incoming signal with the local oscillator signal. One of its output signals becomes the first IF.

**First intermediate frequency amplifier** Is tuned to the IF frequency of 10.7 MHz from the first mixer.

**Second local oscillator** Produces a signal at 455 kHz above the frequency of the first IF signal (11.155 MHz in this case).

**Second mixer** Mixes the first IF signal with the second local oscillator signal to produce the second IF signal.

**Second intermediate frequency amplifier** Is tuned to the second IF frequency of 455 kHz.

**Limiter** Removes all noise or amplitude variations from the signal.

**Discriminator** Demodulates the FM signal and produces the original audio signal. Similar to the detector stage of an AM receiver.

**Audio frequency amplifier** Amplifies the AF from the detector.

**Speaker** Reproduces the original audio signal.

**9-6** FM receiver.

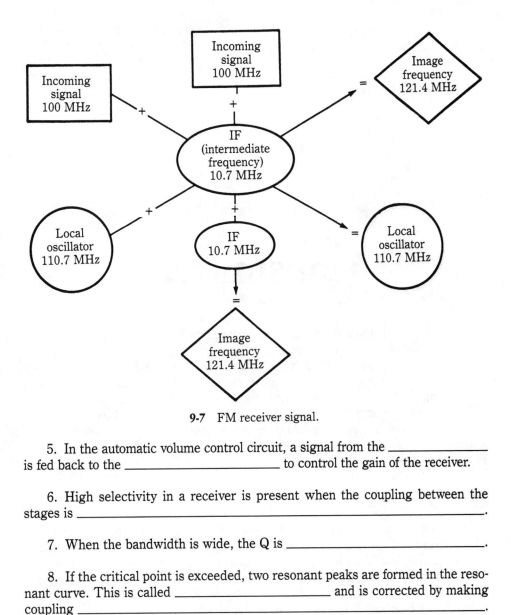

**9-7** FM receiver signal.

5. In the automatic volume control circuit, a signal from the _____ is fed back to the _____ to control the gain of the receiver.

6. High selectivity in a receiver is present when the coupling between the stages is _____.

7. When the bandwidth is wide, the Q is _____.

8. If the critical point is exceeded, two resonant peaks are formed in the resonant curve. This is called _____ and is corrected by making coupling _____.

9. Coupled circuits tend to _____ harmonics. An example is link coupling, which is used in coupling two widely separated circuits.

10. In an FM receiver, the _____ removes all noise or amplitude variations from the signal.

11. In an FM receiver, the _____ demodulates the signal.

# 10
# Transmitters

In the transmitter, intelligence is superimposed onto a radio-frequency carrier wave. In a CW (continuous wave) transmitter, information is placed on the RF wave by means of on-off keying. These transmitters are used in radiotelegraphy. Other types of transmitters superimpose audio intelligence onto the RF wave. They accomplish this by means of either varying the amplitude or the frequency of the carrier wave in accordance with the audio intelligence. The following is a brief overview of a transmitter using AM (amplitude modulation). It will be followed by a more detailed description of the individual components and related topics. See Fig. 10-1.

## Oscillators

An oscillator is an amplifier that is designed to generate a continuous sine wave. The frequency of the sine wave is determined by the *tuned circuit* of the oscillator. In a basic oscillator, the tuned circuit consists of a capacitor and an inductor connected in parallel. The frequency of oscillation depends on the values of capacitance and inductance in the tuned circuit. The oscillations in the tuned circuit are sustained by feedback from the output of the amplifier.

### Crystal-controlled oscillators

In the crystal-controlled oscillator, the tuned circuit is replaced by a quartz crystal. The crystal acts like a tuned circuit. When a sine wave is applied to a quartz crystal, it vibrates mechanically. Likewise, if the crystal is mechanically vibrated, it produces a sine wave. This is known as the *piezoelectric effect*. When a sine wave is applied to the crystal, it vibrates. The vibration in turn produces a sine wave that

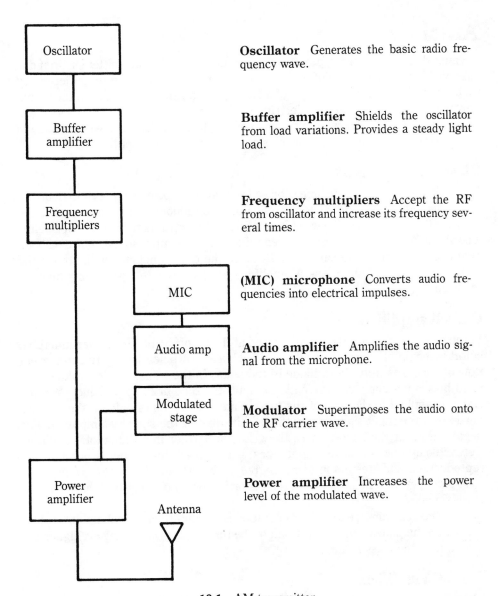

**10-1** AM transmitter.

produces a vibration, etc. The frequency of oscillation is determined primarily by the thickness of the quartz crystal. The crystal-controlled oscillator is the most stable of oscillators. However, the crystal must be shielded from temperature variations, as this will affect its frequency of oscillation. The crystal oscillator, like other types of oscillators, is also sensitive to load variations. Therefore, a buffer amplifier usually follows the oscillator, so that the oscillator frequency does not vary with load changes.

# Amplifiers

Because there are several types of amplifiers used in radio transmitters, a brief discussion of amplifiers is helpful. The primary way of classifying amplifiers is by their class of operation. The class of an amplifier depends upon bias on the amplifier tube or transistor. The bias determines if all or a portion of the input signal will be amplified. In the following discussion of amplifiers, references will be made to the vacuum tube. The same principles apply to transistor amplifiers.

## Class A amplifiers

In the class A amplifier, the tube is biased so that the tube never goes into *cutoff*, or *saturation*. Therefore, the plate current flows continuously. This means that the entire 360° of the input sine wave is amplified. The advantage is excellent fidelity. The amplified signal is a very close reproduction of the input signal. The disadvantage is poor efficiency (about 25%). Because the tube conducts continuously, the plate dissipation is high. Class A amplifiers are used where high fidelity audio is required.

## Class B amplifiers

In the class B amplifier, the tube is biased at the cutoff point. (In the vacuum tube, cutoff is the point where the control grid is biased sufficiently negative to prevent flow of electrons from the cathode to the plate. In the transistor, 0.7 volts of forward bias is the cutoff point.) With no input, the tube does not conduct. When a signal is applied to the grid of the tube, plate current flows only on the positive portions of the input signal. Only the positive half of the signal is amplified. The negative half of the signal drives the grid bias farther into the cutoff condition. Because only one half of the input signal is amplified, the output is not an exact reproduction. This results in some degree of distortion. Two tubes arranged for *push-pull* operation help eliminate the distortion by amplifying both halves of the input signal.

The main advantage of class B operation is higher efficiency (about 50%). This is because the plate conducts only half of the time. The disadvantage is higher distortion than the class A amplifier.

## Class C amplifiers

Class C amplifiers are biased so that only the positive peaks of the input signal are amplified. The class A amplifier amplifies 360° of the input signal. Class B amplifies 180°. Class C amplifies less than 180° of the input signal. The main advantage is high efficiency. Class C amplifiers have the highest efficiency of any amplifiers (about 75%). The disadvantage is poor reproduction of the input signal. For this reason, class C amplifiers are not used to amplify signals that contain audio intelligence. The distortion would be prohibitive. However, class C amplifiers are useful where pure unmodulated sine waves are amplified. They find their use as radio frequency amplifiers and frequency multipliers. The distortion presents no problem because of the flywheel effect of the tank circuit.

## Summary of amplifiers

*Fidelity* is the reproduction quality of the input signal, and *efficiency* relates to the amount of power dissipation required to amplify the signals. Because of their differing characteristics, they have different uses. Table 10-1 summarizes the characteristics of the three basic amplifier classes.

**Table 10-1. Amplifier classes.**

| Class | Fidelity | Efficiency | Use |
|-------|----------|------------|-----|
| A | good | low | audio amp |
| B | fair | medium | audio and RF amp |
| C | poor | high | RF amp and frequency multipliers |

# Frequency multipliers

An oscillator produces a *fundamental* frequency. A harmonic is a whole-number multiple of that fundamental frequency. For example, if the fundamental frequency is 100 kHz, the second harmonic would be 200 kHz, and the seventh harmonic would be 700 kHz. Push-pull operation tends to eliminate even harmonics. In most cases, harmonics should be eliminated because they cause interference to other stations. However, harmonics are essential in frequency multiplier circuits. Any harmonic can be selected by properly tuning the tank circuit. However, higher harmonics have progressively less amplitude.

## Tank circuit

A *tank circuit* is a tuned circuit—a parallel resonant circuit placed at the output of an RF amplifier or frequency multiplier. It consists of a capacitance and inductance in parallel. The resonant frequency is determined by the values of the capacitor and coil. The following formula is used:

$$F = \frac{1}{6.28\sqrt{LC}}$$

where
$L$ = inductance
$C$ = capacitance
$F$ = resonant frequency
$6.28 = 2\pi$

To understand the relationship between $L$ and $C$, you need only substitute some values into the formula. For example, if you wanted to convert a basic RF amplifier into a frequency doubler, you would need to change the values of $L$ or $C$ to cause the tank circuit to resonate at double the frequency—the second harmonic.

For the sake of simplicity, assume that the values of $L$ and $C$ at the fundamental frequency are both 2.

$$\frac{1}{6.28 \times \sqrt{2 \times 2}} = 0.08$$

To get the tank to resonate at twice that frequency, or 0.16 you need to decrease either $L$ or $C$ to one fourth their original value.

$$\frac{1}{6.28 \times \sqrt{2 \times 0.5}} = 0.16$$

Thus, the resonant frequency is double the original.

In frequency multipliers (doublers, triplers, etc.), harmonics are essential. The tube or transistor is biased to produce an output rich harmonics. The output tank circuit is then tuned to resonate at one of the harmonic frequencies. If it has been designed as a frequency doubler, the tank circuit is tuned to the second harmonic. If the tank circuit is tuned to the third harmonic, a frequency tripler is formed. The tank circuit reproduces only the harmonic it is tuned to and attenuates all other harmonics. Class C amplifiers are often used because of the high distortion and harmonic content. Class B amplifiers are also used. In either case, the distorted wave is minimized by the flywheel effect of the tank circuit.

## Flywheel effect

When a class B or class C amplifier is used, considerable distortion is present at the output. This is true because the tube or transistor does not conduct for the full cycle; therefore, the input waveform is only partially reproduced in the output circuit. The output of a class B amplifier is one-half cycle bursts. Class C output contains less than one-half cycle bursts of partial sine waves. These partial sine-wave bursts are fed into a tank circuit that is tuned to resonate at the same frequency at which the tank begins to oscillate. One burst to the tank circuit would cause it to oscillate for a short time, but it would steadily dampen out. However, when a steady stream of bursts from the amplifier nudges the tank circuit, it continues to oscillate. The tank circuit acts like a *flywheel* as it continues to oscillate as long as the partial sine wave bursts continue to nudge it. Thus, the flywheel accepts a distorted, partial sine wave and delivers a pure sine wave at its output. Keep in mind that when the flywheel effect is used, only an unmodulated carrier wave can be used. If audio has been superimposed, severe distortion will result.

# Modulators

The audio signals are converted into electrical impulses in the microphone. They are amplified in the audio amplifiers. The modulator superimposes or injects these audio speech signals onto the radio frequency carrier wave.

## Percentage of modulation

In an amplitude modulation system, the amplitude of the modulating voltage causes the RF carrier wave to vary in amplitude in accordance with that modulating signal. The larger the modulating voltage, the greater is the variation in the amplitude of the RF carrier wave. Greater variation of the carrier equates to a higher percentage of modulation. At 100% modulation, the modulating voltage causes the carrier to vary a full 100%. In other words, the carrier varies from zero to its maximum level. If 100% modulation is exceeded (overmodulation), distortion results. See Fig. 10-2.

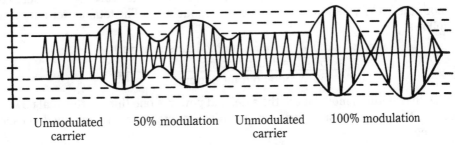

Unmodulated      50% modulation     Unmodulated      100% modulation
carrier                              carrier

**10-2**   Modulation envelope.

## Overmodulation

When 100% modulation is exceeded, distortion, splatter and harmonics are produced. Distortion of the actual modulation envelope occurs, reducing the fidelity of the audio signal. Undesired harmonics can interfere with other stations. Additional sets of sidebands are also created. This *splatter* increases the occupied bandwidth of the signal.

## Power in the sidebands

When the modulating signal mixes with the carrier wave, the following signals are present:
1. The original carrier frequency
2. The modulating frequency
3. Upper sideband (the sum of the two originals)
4. Lower sideband (the difference between the two originals)

The power contained in the sidebands can be calculated as follows:

$$P_{sb} = \frac{m^2}{2} \times P_c$$

where
    $m$ = the percent of modulation divided by 100
    $P_c$ = the power contained in the unmodulated carrier
    $P_{sb}$ = the power in the sidebands

    *Example*   What is the power in the sidebands where the carrier power is 300 W and the modulation is 80 percent?

$$P_{sb} = \frac{0.8^2}{2} \times 300$$

$$= 0.32 \times 300 = 96 \text{ W}$$

*Example*   What is the power in the sidebands where the carrier power is 100 W and the modulation is 100%?

$$P_{sb} = \frac{1^2}{2} \times 100$$

$$= 0.5 \times 100 = 50 \text{ W}$$

In the example, note that at 100% modulation, the modulator stage of the transmitter adds the 50 W when the 100-W carrier is modulated. The total power output is as follows:

$$
\begin{array}{l}
100 \text{ W (power of unmodulated carrier)} \\
+ \;\; 50 \text{ W (sideband power)} \\
\hline
150 \text{ W (total output power)}
\end{array}
$$

Notice that at 100% modulation, the sideband power is one third of the total power. Because there are two sidebands, one sixth of the total power is contained in each sideband.

# Matching networks

A matching network is needed to match the impedance of the final amplifier to the impedance of the transmission line. The most commonly used network is the *pi network*. In addition to impedance matching, the pi network offers excellent harmonic attenuation. A diagram of a pi-network circuit is shown in Fig. 10-3.

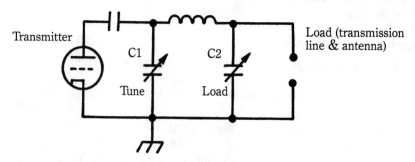

Notice that the capacitor nearer the transmitter is for tuning. The capacitor nearer the antenna (load) is for loading.

**10-3**   Pi network.

Another example of a matching circuit is shown in Fig. 10-4. In this case, C1 is part of a series resonant circuit. This capacitor is used to *tune* the resonant circuit to the output frequency of the final amplifier. When properly tuned to resonance, maximum power transfer is accomplished. C2 is used for *loading*, or matching the output of the transmitter to the load (transmission line and antenna).

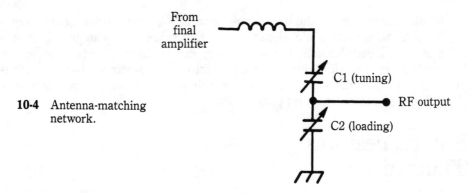

**10-4** Antenna-matching network.

# Parasitics

*Parasitics* are spurious oscillations at undesired frequencies. They result in power losses, reduction of transmitter efficiency, distortion of the output wave, and possible interference with other stations. Although the most likely place for parasitics is in the final RF amplifier, they can occur in any other circuit. The suspected circuits must be isolated and tested independently. The following might be necessary for eliminating parasitics:

- Shielding of leads
- Adding RF chokes
- Noninductive resistors
- Neutralization

# Neutralization

*Neutralization* is a method of eliminating parasitic oscillations in an RF amplifier. The need for neutralization increases as the frequency of operation increases. At UHF frequencies and above, it is essential. Neutralization is accomplished as follows:

1. Remove the plate voltage (B+) from the stage being neutralized. Leave filament voltage on.
2. Tune the preceding stage so that a signal is fed to the stage being neutralized. If this excitation is removed and the following stage still generates RF in its plate tank circuit, neutralization is indicated.
3. Tune the stage being neutralized to resonance while monitoring the RF in the plate tank circuit. Monitoring is accomplished by loosely coupling one of the following devices to the tank circuit:
   a. Neon bulb
   b. Flashlight bulb connected to a loop of wire
   c. Thermocouple ammeter
   d. Grid-dip meter

4. Minimize the RF in the tank circuit by adjusting the neutralizing capacitors.

Neutralization is complete when the net capacitance from plate to grid is canceled to zero. Neutralization is less necessary in tetrode and pentode amplifiers because the screen grid reduces the interelectrode capacitance. The grounded-grid amplifier does not usually require neutralization because the grid acts as a shield between the cathode and plate.

# Study questions
# Transmitters

1. The _____ shields the oscillator from load variations.

2. The _____ superimposes audio on an RF carrier.

3. The frequency of a sine wave is determined by the _____ of the oscillator.

4. _____ is a commonly used crystal in crystal-controlled oscillators.

5. Crystal oscillators are sensitive to _____.

6. The frequency of a crystal is determined by its _____.

7. The oscillating frequency of a crystal can be increased by adding a capacitor in _____. (See *crystal* in glossary).

8. Class A amplifiers have excellent _____ but poor _____.

9. Because of their high harmonic content, class C amplifiers are often used as _____.

10. Which type of amplifier has the greatest efficiency? _____.

11. The _____ effect of the tank circuit tends to remove the distortions from class B or class C RF amplifiers.

12. A harmonic is a _____ of the fundamental frequency.

13. The seventh harmonic of 2182 kHz is _____.

14. To double the resonant frequency of a tank circuit, the value of capacitance or inductance must be _____.

15. The tank circuit of a frequency tripler is tuned to the _____ harmonic.

16. If _____ percent modulation is exceeded, distortion will result.

17. At 100% modulation, the modulator inserts _____ percent of the power.

18. What is the formula for determining power in the sidebands?

19. How much power is contained in the two sidebands at 100% modulation if the power of the unmodulated carrier wave is 1000 W?

20. How much power is contained in the upper sideband if the power of the unmodulated wave is 100 W? _____. What percent of the total power is in the single sideband? _____

21. The pi network is used for _____ the final amplifier to the transmission line.

22. The variable capacitor nearer the transmitter is used for _____ and the variable capacitor nearer the load is used for _____.

23. Parasitic oscillations occur in what circuits?

24. Neutralization is a method of eliminating _____ oscillations.

25. How can they be reduced?

26. The need for neutralization increases as the frequency _____.

27. The _____ voltage is removed on the stage being neutralized.

28. What indicators are used to monitor the RF in the plate tank circuit?
    a. _____
    b. _____
    c. _____
    d. _____
29. Neutralization is complete when the net capacitance from plate to grid is

_____.

30. Neutralization is required in _____ amplifiers but usually not required in _____ amplifiers.

31. A pre-emphasis circuit can be found in an FM _____ while a de-emphasis circuit is found in an FM _____.

32. The pre-emphasis circuit increases the signal-to-noise ratio by selectively attenuating _____ audio frequencies while passing _____ audio frequencies.

# 11
# Bandwidth of emission

Bandwidth of emission (bandwidth) includes all frequencies transmitted with power levels above a certain percent of the total radiated power (above 0.25%). It consists of the carrier, sidebands, and harmonics and contains 99% of the total radiated power.

## Bandwidth of emission in AM transmitters

Amplitude-modulated telephony with double sidebands and full carrier has been designated A3E. It consists of the following components:

**Carrier** The radio frequency wave that carries the audio information. The carrier frequency is the output frequency of the transmitter. The audio information is superimposed (modulated) onto the carrier. The modulation process produces sum and difference frequencies (sidebands) that occupy space above and below the center frequency of the carrier.

**Upper sideband** A band of frequencies above the center frequency of the carrier. The upper sideband is equal to the carrier frequency plus the modulating frequency (the sum).

**Lower sideband** A band of frequencies below the center frequency of the carrier. The lower sideband is equal to the carrier frequency minus the modulating frequency (the difference).

**Explanation** Whenever two frequencies are mixed together, a complex waveform results. The new waveform consists of the two original frequencies plus the sum and the difference frequencies.

*Example* Figure 11-1 illustrates the various frequencies present when a 100-kHz carrier is modulated with a 1-kHz tone. The following frequencies would be present in the complex waveform:

1. 100 kHz (Carrier frequency)
2. 1 kHz (Modulating tone)
3. 101 kHz (The sum of the above frequencies)
4. 99 kHz (The difference of the above frequencies)

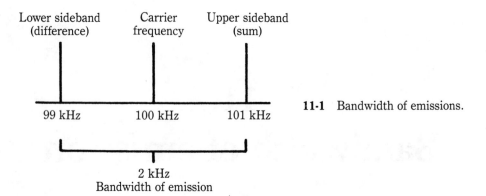

11-1    Bandwidth of emissions.

Because the modulation tone is 1 kHz above and below the carrier frequency, the bandwidth of emission would be 2 kHz. A strong second harmonic of the modulating tone would be equal to twice the original tone, or 2 kHz. That would make the bandwidth of emission equal to 4 kHz. Figure 11-2 illustrates this.

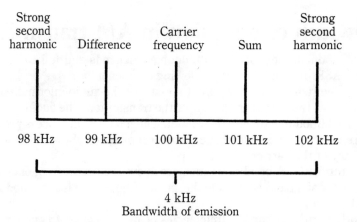

**11-2**    Bandwidth of emission with a strong second harmonic.

If a complex modulating wave is used instead of a pure tone, simply use the highest frequency component of that wave to calculate the bandwidth.

*Example*  If the frequency range of the modulating signal extends from 100 Hz to 2.5 kHz, the higher frequency (2.5 kHz) is used. In this case, the upper sideband would extend to 2.5 kHz above the carrier frequency. The lower sideband would extend to 2.5 kHz below the carrier frequency. The total bandwidth, including the upper and lower sidebands is 5 kHz.

*HINT*: To calculate the bandwidth of an A3E double-sideband AM emission:
- Disregard the carrier frequency
- Bandwidth is equal to *F* times 2

where

*F* = The frequency of the modulating tone or the highest frequency component of a voice-modulating waveform.

2 refers to both sidebands (upper and lower).

## Single-sideband suppressed carrier (SSSC)

This type of emission has been designated J3E and consists of only one sideband. Like A3E, this emission has a carrier frequency, but the carrier is suppressed before the signal is transmitted. Suppression must be at least 40 dB below peak envelope power. In addition to the carrier suppression, one sideband has been filtered out. Because either the upper or lower sideband has been filtered out, this signal occupies one half the bandwidth of an A3E emission. Because of a smaller occupied bandwidth, twice as many stations can share a portion of the frequency spectrum (Fig. 11-3). In the case where the carrier frequency is 100 kHz and the modulating tone is 1 kHz, the J3E transmission will have a bandwidth of 1 kHz, the frequency of the modulating tone.

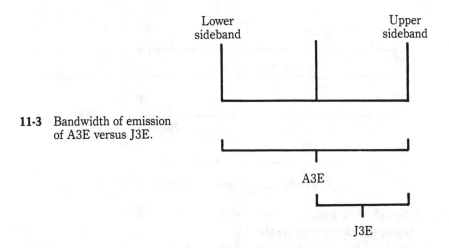

**11-3**  Bandwidth of emission of A3E versus J3E.

*HINT*: To calculate the bandwidth of a J3E SSSC emission:
- Disregard the carrier frequency
- Bandwidth is equal to the frequency of the modulating tone or the highest frequency component of the modulating signal.

# Bandwidth of emission in FM transmitters

The type of emission has been designation F3E and consists of the following components:

**Carrier frequency**  The output frequency of the transmitter.

**Upper sidebands**  A large number of sidebands above the carrier frequency.

**Lower sidebands**  A large number of sidebands below the carrier frequency.

*Explanation*  In AM emission, two sidebands are formed. In FM emissions, many sidebands are formed. In AM, the sideband bandwidth is equal to the frequency of a modulating tone. In FM, the sidebands are spaced apart by the frequency of a modulating tone. The FM bandwidth depends upon the number of significant sidebands generated.

*Example*  If an FM station is modulated with a tone of 5 kHz and there are 8 significant sidebands generated, the total bandwidth would be:

$$8 \times 5 \times 2 = 80 \text{ kHz}$$

In the above case, the deviation above carrier frequency would be 40 kHz. The deviation below carrier frequency would also be 40 kHz. Therefore, the deviation would be 40 kHz from the center. The deviation occurs above and below center frequency. For a deviation of 40 kHz, the bandwidth of emission would be 80 kHz, as shown in Fig. 11-4.

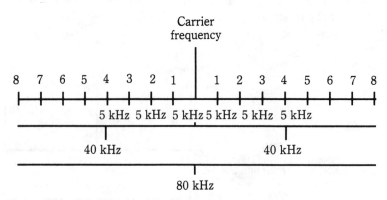

Total bandwidth of emission

**11-4**  Bandwidth of emission of FM transmission.

*HINT*: To calculate bandwidth of an F3E FM emission:
- Disregard the carrier frequency
- Bandwidth is equal to # times *F* times 2

where
   # = The number of significant sidebands
   *F* = The frequency of the modulating tone
   2 refers to the two directions of swing from center frequency

# Study questions
# Bandwidth of emission

1. _____ is all frequencies transmitted with power levels above a certain percent (0.25%) of the total radiated power.

2. Bandwidth of emission contains _____,
_____, and _____.

3. Amplitude modulation (A3E) has the following components:

    a. _____

    b. _____

    c. _____

4. When two frequencies are mixed together (heterodyned), the complex waveform that results consists of the following:

    a. _____

    b. _____

    c. _____

    d. _____

5. A 200 kHz carrier is modulated with a 2 kHz tone.

    a. what is the bandwidth in A3E transmission?

    b. What is the bandwidth in J3E transmission?

6. How is the bandwidth calculated in A3E double sideband emission?

    a. _____

    b. _____

7. In single sideband suppressed carrier (SSSC) transmission, the _____ has been suppressed and one of the _____ has been filtered out.

8. How is bandwidth calculated in F3E (FM telephony) transmission?

    a. _____

    b. _____

9. Determine the bandwidth of an FM carrier wave when it is modulated with a 10 kHz tone. Assume there are nine significant sidebands. _____

10. If the bandwidth of an A3E transmission is 100 kHz, what would the bandwidth be if the transmission was changed to J3E? _____

# 12
# Antennas

The antenna is an integral part of the transmitting system as well as the receiving system. All antennas have one thing in common—to operate at optimum efficiency they must resonate at the frequency of the radio transmission or reception. As a rule, if an antenna is a good transmitting antenna, it is also a good receiving antenna for that same frequency. The reverse is also true.

## Resonant frequency of antennas

The resonant frequency of an antenna is primarily determined by the physical length of the antenna. The relationship between frequency and wavelength must be kept in mind when discussing antennas. That is, as the wavelength increases, the frequency decreases, and vice versa. As the physical length of an antenna is increased, the wavelength is increased, and the frequency is decreased. The resonant frequency of an antenna can be changed by adding an inductor or capacitor in series with the antenna. Table 12-1 summarizes these relationships.

As the frequency increases, the wavelength decreases. The following formulas summarize their relationships.

$$\text{wavelength} = \frac{300}{\text{frequency}}$$

$$\text{frequency} = \frac{300}{\text{wavelength}}$$

where

Frequency is in megahertz
Wavelength is in meters

**Table 12-1. Factors affecting
the resonant frequency of an antenna.**

| To increase resonant frequency | To decrease resonant frequency |
|---|---|
| Make antenna shorter. | Make antenna longer. |
| Add a series capacitor. (A capacitor shortens the antenna) | Add a series inductor. (An inductor lengthens the antenna) |

In the formulas, the speed of light is represented by the 300. The speed of light is 300,000,000 meters per second. One megacycle is 1,000,000 Hz. To simplify the calculations, the six zeros have been canceled out, leaving 300 for the speed of light.

*Example* How are the frequency and wavelength of a 4-MHz radio wave related?

$$\text{wavelength} = \frac{300}{4 \text{ MHz}} = 75 \text{ m}$$

$$\text{frequency} = \frac{300}{75 \text{ m}} = 4 \text{ MHz}$$

It might be necessary to convert to centimeters. To convert, multiply meters by 100.

*Example* What is the wavelength of a 500-MHz radio wave?

$$\text{wavelength} = \frac{300}{500} = 0.6 \text{ m}$$

or

$$0.6 \times 100 = 60 \text{ cm}$$

# Hertz antenna

The Hertz antenna is more commonly called a one-half wave dipole antenna. One-half wavelength is the minimum length of an antenna designed to resonate at a given frequency. The Hertz antenna is fed at its center. Each of the two legs extends one-quarter wavelength to either side of the center. The Hertz antenna is usually horizontally polarized (placed horizontal to the ground).

## Voltage, current, and impedance relationships in a Hertz antenna

In a one-half wave Hertz antenna, the voltage and current are out of phase. The current is minimum at the ends and maximum in the center. The voltage is minimum at the center and maximum at the ends. The impedance is minimum at the center (about 72 Ω) and progressively higher toward the ends. These relationships are summarized in Fig. 12-1.

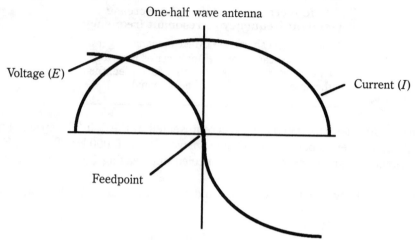

Center: Current (*I*) is maximum and voltage (*E*) is minimum.
Ends: Current is minimum and voltage is maximum.

**12-1**   Current and voltage relationships on a one-half wave antenna.

# Marconi antenna

Antenna is vertically polarized and therefore receives and radiates equally well in all horizontal directions (omnidirectional). A quarter-wavelength radiator is vertically positioned at or near the ground. The second half of the antenna is grounded. There are two methods of feeding such an antenna:

### Series-fed Marconi

A quarter-wave section of the antenna is insulated from the ground and elevated to the desired height above the ground. The second half of the antenna is in the form of several radials—quarter-wave conductors that extend perpendicularly from the base of the antenna. This is known as a *ground-plane antenna.*

### Shunt-fed Marconi

The antenna is not insulated from the ground. Therefore, the dc resistance to ground is zero. Impedance matching is accomplished by tapping the antenna at the desired point above ground. This is possible because the impedance is very low at ground level but steadily increases to a maximum level at the top.

### Voltage, current, and impedance relationships in a Marconi antenna

In a quarter-wave Marconi antenna, the voltage, current, and impedance are the same as in the one-half wave Hertz antenna. In this case, the Hertz antenna is turned on its end and mounted at ground level at its feed point. The second one-quarter leg is grounded. So the current is still maximum at the feed point (center of the antenna), and the voltage is minimum at the feed point. The impedance is minimum at the feed point, just as it is in the center of the one-half wave Hertz. Figure 12-2 summarizes these relationships. Notice that this one-quarter wave Marconi antenna looks just like a one-half wave Hertz antenna that has been rotated 90° and planted into the ground one-quarter wavelength deep.

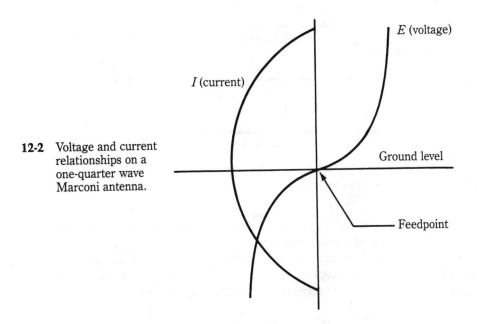

**12-2** Voltage and current relationships on a one-quarter wave Marconi antenna.

# Increasing the gain of an antenna

When more of the radiated signal from the antenna is concentrated in one direction, it is said that the antenna has an increased gain. The antenna becomes less omnidirectional and more directive. Both transmission and reception are improved. Two common methods of increasing the gain and directivity of an antenna are:

**Adding one or more parasitic elements**  A parasitic element is an antenna element that can be slightly longer or slightly shorter than the active one-half wave dipole element. It is placed a fraction of a wavelength from the active element but is not physically connected to it. When the active antenna radiates the signal, a voltage is induced in the parasitic element. It then radiates, thus increasing the directivity of the antenna in one direction. This two-element antenna is more directional than a single-element antenna. As more parasitic elements are added, the gain and directivity of the antenna increase. Additionally, as the forward directivity increases, received signals that enter through the sides or the back of the antenna are attenuated. The amount of attenuation is described in terms of the front-to-side ratio, and front-to-back ratio of the antenna. Figure 12-3 illustrates the effect of parasitic elements.

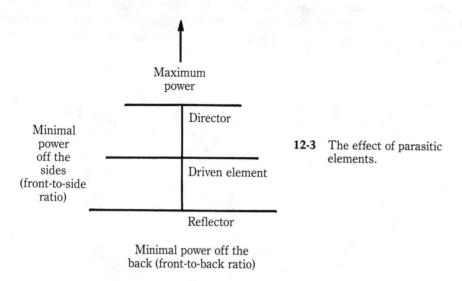

**12-3** The effect of parasitic elements.

**Stacking antennas**   Stacked antennas are two or more antennas that are mounted a fraction of a wavelength apart. They are mounted one above the other if they are horizontally polarized, or side by side if they are vertically polarized. Stacking antennas increases the directivity and gain of an antenna used for transmitting, and increases the reception of a receiving antenna.

# Electromagnetic radiation

When a transmitter sends a radio frequency wave to the antenna, electromagnetic radiation occurs at the antenna. This radiation consists of electric and magnetic lines of force that are 90 degrees out of phase. They then travel through space at the speed of light. The two fields are induced in a receiving antenna. The type of polarization is determined by the electric field of the antenna. The metal element represents this field. Therefore, if the antenna is positioned parallel to the surface of the earth, it is horizontally polarized.

# Study questions
# Antennas

1. The _____ of an antenna is primarily determined by its physical length.

2. To increase the resonant frequency of an antenna, make it _____.

3. A series inductor will _____ the resonant frequency of an antenna.

4. As the frequency increases, the wavelength _____.

5. What is the wavelength of a 100-MHz radio wave in meters?

6. What is the wavelength of a 500-MHz radio wave in centimeters?

7. What is the frequency of a 30-m radio wave?

8. A one-half wave dipole is called a _____ antenna.

9. The Hertz antenna is usually _____ polarized.

10. Draw a sketch of the voltage relationship along a one-half wave Hertz antenna.

11. Draw a sketch of the current relationship along a one-half wave Hertz antenna.

12. What is the voltage and current relationship at the center?

13. What is the voltage and current relationship at the ends?

14. The Marconi antenna is _____ polarized.

15. In the Marconi antenna, one-quarter wave acts as a radiator and the second half is _____.

16. The _____ fed Marconi is insulated from the ground.

17. The _____ fed Marconi is not insulated from the ground. For that reason, the dc resistance to ground is _____.

18. _____ and _____ are two ways to increase the gain of an antenna.

19. Stacking antennas increases the _____ of the antenna and this results in _____ reception.

20. One or more parasitic elements will _____ the directivity and will make the antenna _____ directional.

# 13
# Transmission lines

Transmission lines carry the radio frequency (RF) energy from the transmitter to the antenna, and from the antenna to the receiver. One of the most common types of transmission line is called *coax* (coaxial) cable. Coax has a wire conductor surrounded by an insulating (*dielectric*) material. Around the insulating material is another conductor in the form of a metal braid. These concentric conductors are surrounded by a nonconductive sheath that provides protection from the weather. Some cables are filled with nitrogen, an inert gas that prevents moisture from entering the transmission line.

## The importance of impedance matching

For maximum power transfer between any two electrical circuits, an impedance match is required. That is, the impedance of the load must match the impedance of the source. Figure 13-1 illustrates an ac source with a load resistor. The source could represent a radio transmitter, and the load resistor could represent an antenna. You can determine how much power is transferred from the source to the load as follows:

1. Calculate the current flow ($I$) in the circuit.

$$I = \frac{E}{Z_t}$$

where

$E$ = Voltage of the source
$Z_t$ = Total impedance in the circuit

2. Calculate the power ($P$) delivered to the load.

$$P = I^2 \times Z_L$$

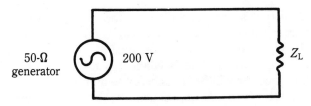

**13-1**   Power transfer between generator and load.

Assume that the source generates 200 V. If the load impedance matches the 50-$\Omega$ source impedance, the power delivered to the load is determined as follows:

$$I = \frac{E}{Z_t}$$

$$= \frac{200}{100}$$

$$= 2 \text{ A}$$

Power to the load $= I^2 Z_L = (2)^2 \times 50 \ \Omega = 200 \text{ W}$

Note that no other value of load resistance will yield greater power transfer to the load. Maximum power transfer occurs when the source and load impedances are matched.

If the load impedance is increased to 75 $\Omega$:

$$I = \frac{E}{Z_t}$$

$$= \frac{200}{125}$$

$$= 1.6 \text{ A}$$

Power to the load $= I(2) \ Z_L = (1.6)^2 \times 75 \ \Omega = 192 \text{ W}$

If the load impedance is decreased to 25 $\Omega$:

$$I = \frac{E}{Z_t}$$

$$= \frac{200}{75}$$

$$= 2.7 \text{ A}$$

Power to the load $= I^2 Z_L = (2.7)^2 \times 25 \ \Omega = 182 \text{ W}$

When the source impedance matches the load impedance, maximum power is delivered to the load. If the load impedance is either increased or decreased, the power transfer decreases.

# Power transfer in transmission lines

Transmission lines have a characteristic impedance (surge impedance). This is determined by the physical dimensions of the line and the characteristics of the dielectric material that separates the conductors. To achieve maximum power transfer between a transmission line and an antenna, the impedance of the antenna must be equal to the surge impedance of the line that feeds it. If, for example, a 50-Ω transmission line is connected to a properly resonated 50-Ω antenna, maximum power will be transferred to and radiated from the antenna. Figure 13-2 illustrates how the impedance of the transmitter must match the surge impedance of the transmission fine to achieve maximum power transfer from the transmitter to the line. The line in turn is connected to an antenna with the same impedance as the line. In this case, all the forward power from the transmitter is absorbed by and radiated by the antenna. In such a properly matched transmission line, the voltage and current will be the same at all points along the line. The rms voltage along this line is equal to the product of the surge impedance of the transmission line times the current flowing through the line ($E = IR$). This, of course, is the optimum arrangement.

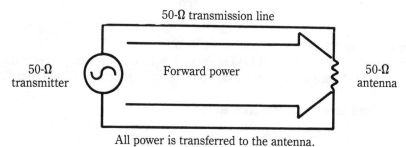

50-Ω transmission line

50-Ω transmitter

Forward power

50-Ω antenna

All power is transferred to the antenna.

**13-2** Power transfer in a matched line.

If the same 50-Ω line is connected to a 100-Ω antenna, an impedance *mismatch* is present. Less than maximum power is transferred to the antenna. A portion of the RF power that reaches the antenna is reflected back down the line. Figure 13-3 illustrates this. This reflected wave interacts with the forward wave and causes *standing waves* in the transmission line. These standing waves correspond to maximum and minimum points of voltage and current along the line. The impedance mismatch results in power loss because not all of the power is transferred to the antenna. If the standing waves are high enough, arcing might occur in the line. The SWR (*standing wave ratio*) is the relationship of the maximums to minimums.

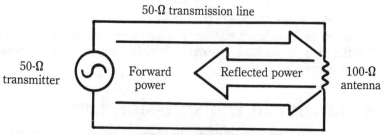

50-Ω transmission line

50-Ω transmitter

Forward power

Reflected power

100-Ω antenna

This impedance mismatch causes reflected power.

**13-3** Power transfer in an unmatched line.

SWR is expressed as follows:

$$\text{Standing wave ratio} = \frac{\text{maximum voltage}}{\text{minimum voltage}} \text{ or } \frac{\text{maximum current}}{\text{minimum current}}$$

SWR is always expressed as the ratio of a number to the number 1, such as 3:1, 2:1, or 1:1. An SWR of 1:1 is the most desirable. Another method of calculating the SWR is by dividing the transmission line impedance and the antenna impedance in such a way that a positive number results. For example, if the surge impedance is 50 Ω and the antenna impedance is also 50 Ω, the SWR would be 1:1. If the above antenna is replaced with a 100-Ω antenna, the SWR would be 100/50 or 2:1. If the antenna impedance were changed to 40 Ω, the SWR would be 50/40 or 1.25:1.

The greater the impedance mismatch, the higher the SWR. The higher the SWR, the greater the power loss. To maintain the lowest possible SWR, and the maximum transfer of power, the surge impedance of the transmission line must be equal to the impedance of the antenna.

A useful device for determining the relationship between forward and reflected power is the *directional wattmeter*. With it, the actual power delivered to the antenna or dummy load is determined. For example, if a directional wattmeter indicates 30 W of forward power and 5 W of reflected power, the power delivered to the antenna is only 25 W. Another way of looking at it is if the directional wattmeter indicates 45 W of forward power and 5 W of reflected power, the actual transmitter output power is 40 W. To determine true forward power, subtract the reflected power from the forward power as in the following:

$$\text{True forward power} = \text{forward power} - \text{reflected power}$$

# Characteristics of quarter-wave sections of transmission lines

**Termination of transmission lines**   By terminating transmission lines in different manners, impedance is inverted or matched.

**Quarter-wave line terminated in resistance**   As discussed above, when the line is terminated in a resistance that equals the characteristic impedance of the line, no standing waves are formed. Therefore, the voltage is the same level at all points along the line. The line is called *flat*. Its rms voltage is the product of the impedance (resistance) times the line current ($E = I \times R$). It acts like an infinitely long transmission line. It is as though the RF energy never reaches the end of the line, because there are no reflections. No reflections occur because all the RF energy (not considering line losses) is absorbed by the antenna. The input impedance of this line is equal to its output impedance.

**Quarter-wave line terminated in a short**   When the end of the transmission line is shorted, standing waves are produced. The voltage and current relationships are summarized in Table 13-1.

**Quarter-wave line terminated in an open circuit**   Standing waves are also produced in this case. The voltage and current relationships are summarized in Table 13-2.

**Table 13-1. Characteristics of
a quarter-wave line terminated in a short.**

|  | Voltage (*E*) | Current (*I*) | Impedance (*Z*) |
|---|---|---|---|
| *Input* | high | low | high |
| *Output* | low | high | low |

**Table 13-2. Characteristics of
a quarter-wave line terminated in an open.**

|  | Voltage (*E*) | Current (*I*) | Impedance (*Z*) |
|---|---|---|---|
| *Input* | low | high | low |
| *Output* | high | low | high |

Figure 13-4 illustrates the above characteristics of transmission lines with terminations that are open versus shorted. Additionally, it gives the characteristics of series and parallel resonant circuits. The diagram can be reconstructed in just a few seconds as follows:

1. Draw two parallel horizontal lines of equal length.
2. Draw an X that connects all four ends.
3. Draw a vertical line at the far end, indicating a short.
4. Identify the points of maximum current. Because maximum current flows through a short circuit, identify the upper right sections as (*I*). Note that the *E* (voltage) and *Z* (impedance) are minimum at that point.
5. The remaining points can then be quickly identified.
6. The above steps identify the characteristics of a transmission line terminated in a short. To determine the characteristics of a transmission line terminated in an open circuit, simply reverse the points. For example, the current (*I*) would be minimum and the *EZ* would be maximum at the open end.

# Resonance

As the frequency is increased, the reactance of a capacitor (capacitive reactance) steadily decreases, while the reactance of an inductor (inductive reactance) steadily increases. Therefore, if an inductor and capacitor are connected together in a circuit that is subjected to various frequencies, a frequency can be reached where the reactances are equal. This is called the *resonant frequency*. At resonance, the capacitive reactance equals the inductive reactance and they cancel each other, leaving only pure resistance in the circuit. Resonant circuits are in the form of series or parallel circuits. The parallel resonant circuit is the tank circuit that was

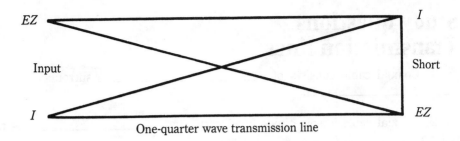

One-quarter wave transmission line

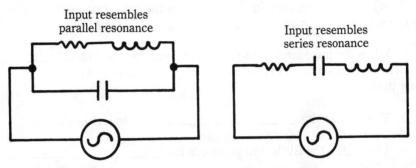

**13-4** Input and output characteristics of a one-quarter wave shorted line.

previously discussed. Series and parallel resonant circuits have different characteristics. The reason resonance is discussed in this chapter is because the preceding
diagram of the quarter-wave transmission line terminated in a short circuit gives all
the characteristics of both types of resonant circuits. The diagram can be sketched
in a few seconds and used as a reference as follows:

- The input end (portion to the left) of the quarter-wave transmission line that
  is terminated in a short has the following characteristics of a parallel reso-
  nant circuit:
  - ~ Maximum voltage ($E$)
  - ~ Maximum impedance ($Z$)
  - ~ Minimum current ($I$)
- The termination end of the transmission line resembles a series resonant
  circuit as follows:
  - ~ Maximum current ($I$)
  - ~ Minimum voltage ($E$)
  - ~ Minimum impedance ($Z$)

Notice that the impedance ($Z$) and the voltage ($E$) are easy (E-Z) to remember
because they always accompany one another ($EZ$) on the diagram. A good place to
start is to identify the point of maximum current in the diagram. A short circuit
would logically permit a maximum current to flow. Maximum current is therefore
found at the shorted end of Fig. 13-4.

# Study questions
# Transmission lines

1. Coaxial cable consists of a _____ surrounded by a _____.

2. Coaxial cable is sometimes filled with _____ gas to reduce _____ in the line.

3. When the impedance in the antenna is not the same as the impedance of the transmission line, an _____ exists.

4. When the impedance is properly matched, there will be no _____ waves in the transmission line and the voltage and current will _____ along the line.

5. The greater the _____, the higher the SWR.

6. The higher the SWR, the greater is the _____.

7. A transmission line that is terminated in the proper resistance will have no _____. When the input impedance equals the output impedance, the line acts like an _____ transmission line.

8. Construct the diagram that gives the relationships of voltage, current, and impedance in one-quarter wave section of transmission line.

9. At resonance, the capacitive reactance is equal to the _____, and they cancel one another out, leaving pure resistance.

10. With the diagram from question number 8, list the characteristics of series and parallel resonant circuits.

11. For maximum power dissipation, $R_L$ should be: _____ (Fig. 13-5).

30 Ω      20 Ω

$R_L$      **13-5**   Impedance matching.

# 14

# Effective radiated power (ERP)

Effective radiated power (ERP) is the actual power that leaves the antenna. To calculate ERP, take into account the following:
- Transmitter efficiency
- Transmission line loss
- Equipment loss
- Antenna gain

## Calculating ERP

Calculate ERP by multiplying the total system gain (actually, the power change) by the transmitter output power. This is illustrated in Fig. 14-1. Here are the steps:

1.  Determine system loss. For example, if the transmission line has a 3 dB loss and a wattmeter in the line has 1 dB loss, the total system loss is 4 dB.
2.  Subtract system loss from antenna gain. If the antenna, in this example, has a forward gain of 14 dB, the total system gain would be 14 dB – 4 dB = 10 dB.
3.  Convert decibels to power change. Various simple methods of converting to power change are shown in this chapter. For the sake of this example, a 10 dB gain results in a power change of 10.
4.  Multiply power change by transmitter power. Assume that, in this example, the transmitter output power is 100 W. The ERP equals power change times transmitter output power, or $10 \times 100 = 1000$ W.

Now take a look at the individual considerations:

### Transmitter efficiency

In order to calculate the effective radiated power of the system, you must know the actual output power of the transmitter. Transmitter output power is the product of

**165**

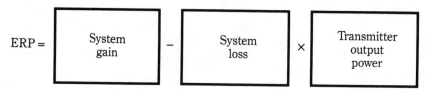

**14-1**   ERP (effective radiated power).

plate current and plate voltage at the final power-amplifier stage. Multiply that figure by an efficiency factor. If the plate voltage is 600 V, the plate current is 700 mA, and the final amplifier efficiency is 75%, the transmitter output power would be $600 \times 0.7 \times 0.75 = 315$ W.

## Transmission line loss

Coaxial cable is rated for decibel loss per 100 ft. of length. The line loss increases as the operating frequency increases. For example, RG-58/U has a cable loss of 3 dB per 100 ft. at a frequency of 50 MHz. If you increase the operating frequency to 420 MHz, the loss increases to 15 dB per 100 ft. Figure 14-2 illustrates the linear nature of line loss. Please note that if the loss is 3 dB per 100 ft., a 200 ft. transmission line would produce a 6 dB loss. If 150 ft. of the coax is used, the loss would be 4.5 dB. With this concept understood, you can determine the dB line loss for any length of cable.

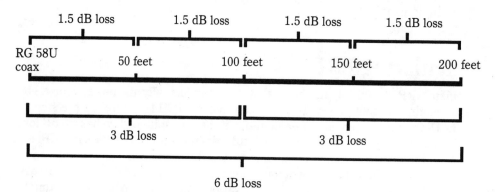

**14-2**   Power loss in a transmission line.

## Equipment loss

Whenever radio frequency energy is passed through an electrical device, a certain amount of signal loss is sustained. The equipment might include: an SWR bridge, a wattmeter, a duplexer, etc. Equipment loss is rated in decibels (dB). The equipment loss plus the transmission line loss constitute the system loss that is used to calculate ERP.

### Antenna gain

The transmitter output power is attenuated by line loss and equipment loss. The remaining power enters the antenna where it is amplified and concentrated in one direction. Remember that the antenna gain can be increased by adding additional parasitic elements and stacking antennas. Antenna gain is measured in decibels.

# Calculating power change

Once the gain is balanced against the loss, the remaining figure is the total system gain. This decibel value must be converted into a power change value. *Power change* describes how much the power level increases, for a gain, or decreases, for an attenuation. The product of the power change and the transmitter output power is the ERP. You will read about two simple methods for calculating power change. One method requires a calculator with a $10^X$ function. The other method requires only a pencil and paper.

### Pencil method for calculating power change

The method is for field use or whenever you don't have a scientific calculator. Remember three keys in order to do the pencil method. First, 1 dB gives a power change of 1.26. Second, 3 dB gives a power change of 2. Third is that you must know how much power change is associated with even powers of 10 dB. This is the 10 - 20 - 30 rule. Table 14-1 summarizes the rules. The power change for 10 dB, 20 dB, 30 dB, etc., will always begin with a 1 and end with one or more zeros. The first number in the dB value tells you how many zeros to add to the 1. For example, a 10-dB gain equates to a power change of 10. A 20-dB gain results in a power change of 100. A 30-dB gain would have a power change with three zeros, or 1000. Through further study of Table 14-1, notice that the power change for a signal loss, or attenuation, is calculated in much the same manner. With attenuation, the 1 with its zeros is divided into 1. This is accomplished by pressing the 1/X key on the calculator. The FCC expects students to be able to answer decibel questions for gain and attenuation, with answers in decimals or fractions. Careful study of Table 14-1 is important.

**Table 14-1. The 10-20-30 Rule for dB calculation.**

| Power change | Gain | Attenuation | Power change | |
|---|---|---|---|---|
| 10x | 10 dB | − 10 dB | 1/10x | (0.1x) |
| 100x | 20 dB | − 20 dB | 1/100x | (0.01x) |
| 1000x | 30 dB | − 30 dB | 1/1000x | (0.001x) |
| 10000x | 40 dB | − 40 dB | 1/10000x | (0.0001x) |

Once you know the above three keys, you can quickly solve decibel problems with only a pencil. First, draw a horizontal line across a sheet of paper. Above the line, break down the decibels into the components that you know. Figure 14-3 illustrates how a 14 dB gain is converted into a power change figure. Notice that the 14 is broken down into 10 dB, 3 dB, and 1 dB, all of which the reader should recognize. Below the line, directly under the dB values, write the power change associated with the decibel. The numbers above the line are added together to equal the total dB value. The power change numbers, below the line are multiplied to determine the total power change, which in this case is 25.2.

| Decibels (dB) | 10 dB | + | 3 dB | + | 1 dB | = | 14 dB |
|---|---|---|---|---|---|---|---|
| Power change | 10 | × | 2 | × | 1.26 | = | 25.2 |

**14-3**   Pencil method for decibel calculations.

Another example is converting 7 dB to a power change. By breaking down 7 dB, you have $3 + 3 + 1 = 7$ dB. The power change is $2 \times 2 \times 1.26 = 5.04$

Once you determine the power change figure, simply multiply it by the transmitter output power to find the effective radiated power. For example, if the transmitter power is 100 W and the total system gain is 14 dB, the effective radiated power would be $100 \times 25.2 = 2520$ W.

## Calculator method for power change

When the first and second editions of this study guide were prepared, the FCC was only asking questions that were 10 dB, 20 dB, 30 dB, etc. Recently, questions have appeared that required the test taker to know power changes for odd decibel values that do not fit into the 10 - 20 - 30 rule. This calculator method will enable you to calculate the power change for any decibel value nearly instantly.

This method requires a scientific calculator. The calculator must have a LOG key and an $10^X$ (sometimes called *antilog*) key. By working with the dB log equation:

$$dB = 10 \ Log \ p/p$$

By working with the numbers you can simplify the calculations. The new formula is:

$$10^{\frac{dB}{10}}$$

Please note that the above formula applies only to power calculations. If you are working with voltage of current, it must be changed to:

$$10^{\frac{dB}{20}}$$

If your calculator does not have a $10^X$ key, use the antilog key. You might have to press the inverse function or 2nd function key first to activate it.

Enough preliminaries! See just how easily it works by working several examples.

*Example* What is the power change in an amplifier with a 20-dB gain? 20 divided by 10 equals 2. With 2 on your calculator display, press $10^X$ or whatever is needed for antilog. The answer is 100. But you could have solved that one in your head, using the 10 - 20 - 30 method. Try another one.

*Example* What is the power change in an antenna with a 3-dB gain? On your calculator, divide 3 by 10, then press the $10^X$ key. The answer is 1.995, or roughly 2. In other words, if your antenna has a 3-dB gain, the radiated power out of the antenna is twice that of the power at the input of the antenna.

*Example* What is the power change in a 7-dB amplifier? Divide 7 by 10 and press the $10^X$ or antilog key to get 5.01 displayed on your calculator. In other words, if 10 W is fed into a 7-dB amplifier, you would have 50 W (10 W times 5.01) at the output. You simply calculate the power change, then multiply it by the input power.

*Example* What is the output of a 13-dB amplifier, if 0.2 W is fed into it? Divide 13 by 10 and press $10^X$. The power change is 19.95, or roughly 20. Then you must multiply 20 times 0.2 to get 4 W.

*NOTE*: You can use this same method with voltage or current gain or loss. Instead of dividing by 10, divide by 20. For example, an amplifier having a voltage gain of 30 dB has 0.01 V applied to its input. What is the output voltage? By dividing 30 dB by 20 and pressing $10^X$, your calculator should show 31.6. This is the voltage change. To determine the output voltage, multiply 31.6 by the 0.01 input voltage to get 0.316 V.

In summary, when you are calculating gain, you multiply the input power (of the amplifier or antenna) times the power change number. The FCC test will also have dB loss (or attenuation) problems. The first half (calculating the power change) remains the same. But after you determine that power change figure, you divide instead of multiplying. Look at some examples.

*Example* 100 W is passed through a transmission line with a 3-dB line loss. How much power is present at the other end of the transmission line? You probably remember from a previous question that 3 dB gives a power change of 2. So instead of multiplying—divide the input power by 2, or $100 \div 2 = 50$ W at the other end of the line.

*Example* 100 W passes through a 6-dB attenuator. How many watts appear at the output of the attenuator? Press $6 \div 10 = 10^X$. You should see 3.98 on your calculator. Round it off to 4. Then press $100 \div 4 =$, and you 25 W.

## Converting power change into decibels

You can convert from power change to a decibel value as follows:

$$10 \log (\text{power change})$$

In other words, enter the amount of power change on the calculator. Press the log button. Then multiply by 10 to calculate the decibels. This is true for power changes.

*Example*   If a transmitter increases its ERP from 100 W to 500 W, how much decibel gain has occurred? Enter 5 and press the log key. Multiply the answer by 10 to get 7 dB.

Please note that the pencil method can get you very close to the exact answer. It will work for you in the field or during the FCC test, if you forget your calculator. But it is easier and more accurate to use the $10^X$ method with your calculator. Protect yourself—learn both methods.

# Field strength

An electromagnetic field emanates from the transmitting antenna. This field increases directly with the effective radiated power. Field strength, or field intensity, is measured in millivolts per meter ($\mu$V/m) or microvolts per meter with a field-strength meter.

## Increased distance

The field intensity decreases inversely with the distance from the antenna. If the distance from the transmitter is doubled, the field strength is half. If the distance is tripled, the intensity is one third. And if the distance is increased four times, the field intensity would be reduced to one fourth. Table 14-2 summarizes this.

**Table 14-2. The effect of distance on field strength.**

| Distance from transmitter | Field strength |
|---|---|
| 2x | 1/2 |
| 3x | 1/3 |
| 4x | 1/4 |
| 5x | 1/5 |

## Increased transmitter power

If the transmitter power is doubled, the field strength at a given distance will increase by 1.414, which is the square root of 2. If the transmitter power is increased by a factor of three, the field strength will increase by the square root of 3, which is 1.73 times. If the transmitter power is quadrupled, the field strength will be doubled, which is the square root of 4. Simply determine by what factor the transmitter power has been increased. Take the square root of that number and multiply that times the original field strength to get the new one. For example, a

100-W transmitter produces a 40 $\mu$V/m field strength. If the transmitter is increased to 700 W, or seven times stronger, take the square root of 7. Multiply 2.65, the square root of 7, times 40 $\mu$V, the original field strength, to arrive at a new field strength of 106 $\mu$V. Table 14-3 summarizes this concept.

**Table 14-3. Calculation of field strength when transmitter power is increased.**

| Field strength original ($FS_o$) | Transmitter power original (x) |
|---|---|
| $FS_o \times \sqrt{2}$ | 2x |
| $FS_o \times \sqrt{3}$ | 3x |
| $FS_o \times \sqrt{4}$ | 4x |
| $FS_o \times \sqrt{5}$ | 5x |

## Increased field strength

You can determine the amount of transmitter power change by monitoring the field intensity at a given distance. It is much the same as the process described above, except instead of using square roots, you square the number. For example, if the field strength doubles, you can calculate the increase in transmitter power by squaring the number 2, as illustrated in Table 14-4. The square of 2 is 4, which means the transmitter is 12 times stronger, the transmitter power is 12 times 12, or 144 times stronger to create a 12 times stronger field strength.

**Table 14-4. Calculation of transmitter power change with a known field strength change.**

| Original transmitter power (P) | Original field strength (x) |
|---|---|
| $P \times 2^2$ | 2x |
| $P \times 3^2$ | 3x |
| $P \times 4^2$ | 4x |
| $P \times 12^2$ | 12x |

If you measure a field strength of 20 $\mu$V/m, before a transmitter power increase and 160 $\mu$V/m after the power increase, by what factor was the power increased? By dividing 160 by 20, you see that the field strength was eight times stronger after the power increase. The square of 8 is 64. The transmitter power was 64 times greater to produce the above change in field strength.

# Study questions
# Effective radiated power

1. The field strength, at a given distance, is 0.01 $\mu$V/m. How many times would the transmitter power have to be increased to get a field strength reading of 0.04 $\mu$V/m?

2. The field strength is 50 $\mu$V/m at a given distance. What would the field strength be at twice that distance if the transmitter power was increased by a factor of 9?

3. What is the effective radiated power of the following system: transmitter output power is 100 W; 50 ft. of coax, with a line loss of 3 dB per 100 ft.; wattmeter in the line has a loss of 2 dB; antenna gain is 12 dB.

4. What happens to the line loss in a transmission line, if the operating frequency is increased?

5. What are the power changes for 1 dB, 3 dB, and 10 dB?

# 15

# UHF and above

As the operating frequency is increased, the circuits and circuit components become more critical. For example, vacuum tubes that operate well at 5 MHz will not operate at 900 MHz. Special UHF and above devices have been developed. Resistance, capacitance, and inductance become more critical as the frequency increases. There are special considerations in transmission lines. Even equipment grounding techniques that work well at low frequencies become ineffective at higher frequencies.

## Circuit considerations

The following items must be considered whenever UHF is involved:

**Capacitance** Stray capacitance can present a problem at UHF frequencies. As you will recall from the capacitive reactance formula, the reactance becomes less as the frequency increases. It is important to keep the length of the wire leads as short as possible. A long lead, for example, that runs near the chassis, can have enough capacitance to shunt much of the signal to ground.

**Inductance** Stray inductance from loops or bends in wire leads will result in increased inductive reactance, and therefore, attenuation of the signal. Small amounts of inductance in a capacitor can be sufficient enough to form unplanned resonant circuits that attenuate the signal.

**Resistance** The carbon resistor works well at low frequencies. However, if it is used in the UHF range, it will appear to the circuit as having components of capacitance and inductance.

**Grounding** As the frequency is increased, the RF ground becomes increasingly important. For example, a heavy gauge copper wire to a good ground rod would provide an effective ground at low frequencies. At UHF and above, it is best to get away from wire entirely. *Grounding foil* is a thin sheet of copper that replaces the copper wire. The copper sheet has lower reactance at high frequencies and is therefore preferable.

**Skin effect**   The *skin effect* is the tendency of RF to travel on the outer surface of the conductor rather than through the center. This is because of lower reactance on the surface. Therefore, there is less loss of signal with larger diameter wire leads because of increased surface area. The thin sheet of copper wire (grounding wire) has even more surface area.

# Transmission lines

Transmission line losses also increase at high frequencies. Modifications have been made in coaxial cable to minimize losses. At higher frequencies, waveguides are used.

**Coaxial cable**   To reduce the dielectric loss in coaxial cable, air is often used to replace the dielectric insulating material between the conductors. Nitrogen gas is sometimes inserted into the cable to replace the insulator material. This reduces the dielectric losses and prevents moisture from entering the cable.

**Waveguides**   *Waveguides* are high-frequency transmission lines. They are constructed as either circular or rectangular hollow metallic conducting tubes. Their size corresponds to the wavelength of the frequency they are designed to conduct. Generally, the radius of the waveguide must be equal to or greater than, one third of the wavelength of the operating frequency. Because higher frequencies have shorter wavelengths, smaller waveguides are required. A waveguide offers excellent conduction with low loss at high frequencies. However, the required size would prohibit using a waveguide at low frequencies. Moisture in the waveguide will cause signal attenuation. To reduce moisture, pressurized nitrogen gas is inserted. Long horizontal runs of waveguide should also be avoided because of potential moisture accumulation.

# Vacuum tubes

As the frequency is increased, the electron transit time becomes critical, (the time required for the electrons to travel from the cathode to the plate of the tube) if the frequency is high enough, the transit time can equal the period of the RF sine wave. This results in phase distortions.

**UHF tubes**   Special tubes have been constructed that minimize transit time and, therefore, allow operation at higher frequencies. Other factors like interelectrode capacitance and lead inductance have been reduced by tube construction.

**Magnetron tube**   This tube is a type of diode that must be surrounded by an intense magnetic field. It is used as a self-contained oscillator to generate super high frequency (SHF) signals.

**Traveling-wave tube (TWT)**   The TWT is a UHF-and-above amplifier tube. An electron gun directs a beam of electrons to an anode at the far end of a hollow tube. The tube is surrounded with a wire helix. The input microwave signal is coupled to the helix at the cathode end (closest to the electron gun). As the electrons conduct through the helix, the magnetic field that is generated around the helix interacts with the electron beam to produce bunches of electrons. This bunching, in turn, induces currents in the helix that result in amplification of the microwave signal.

# Ferrite devices

Ferrites are specially constructed rods consisting of various combinations of metallic oxides. When surrounded with a magnetic field, a molecular resonance occurs in the rod. The strength of the magnetic field helps to determine the resonant frequency of the rod. RF energy equal to the resonant frequency of the rod will interact with its molecular resonance, causing absorption of the RF energy.

**Ferrite switches**   If a ferrite is placed within a waveguide, it can be made to act as an RF switch. When the magnetic field is applied to the rod, the RF energy is absorbed. When the magnetic field is inactivated, the RF energy passes attenuated through the rod.

**Isolators**   The isolator is surrounded by a permanent magnetic field. When RF energy enters the rod from one direction, it interacts with the molecular resonance of the rod. The RF is absorbed by the rod, resulting in strong attenuation of the signal. However, if the RF enters the rod from the opposite direction, no interaction or attenuation occurs. It acts like a one-way valve for microwave signals. The isolator has one input port and one output port. Because of their one-way characteristic, isolators are useful for:

- Attenuating reflected waves in an improperly matched transmission line (Fig. 15-1).
- Helping to reduce intermodulation by preventing external signals from entering a transmitter through its antenna (Fig. 15-2). If transmitter 2 is located in close proximity to transmitter 1, its signal can enter the antenna of transmitter 1, travel down the transmission line and interact in the transmitter, causing intermodulation distortion. The isolator prevents this by attenuating the signal from the external transmitter in much the same way as it attenuates reflected waves.

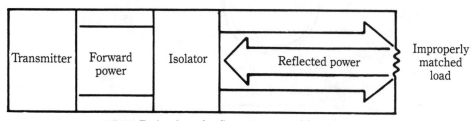

**15-1**   Reduction of reflected waves with an isolator.

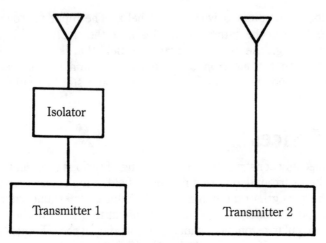

**15-2**   Reduction of intermodulation with an isolator.

**Circulators**   The circulator is similar to the isolator except that it has three or more ports, instead of two. When RF energy enters the circular ferrite device, it becomes circularly polarized. This RF wave will also undergo a phase shift or rotation in its polarization. Remember that if a wave is not properly polarized, it will not propagate. The circulator is so designed that the phase shift causes the wave to exactly match the polarization of the next port, allowing the signal to exit at that port only. If a port is terminated with a reflective load, the wave will bypass that port and appear at the next available port. Figure 15-3 illustrates how a circulator is used as a duplexer to connect a receiver and transmitter to a common antenna.

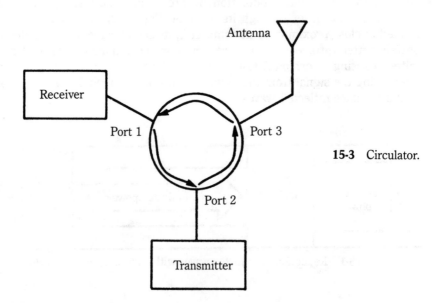

**15-3**   Circulator.

Notice how the received signal from the antenna progresses only to the receiver port. Also notice how the high power RF from the transmitter proceeds only to the antenna. The receiver is isolated from the transmitter. Figure 15-4 shows how one port is terminated in a resistance, converting the circulator into an isolator. The RF energy is totally absorbed by a resistance that matches the transmission line impedance.

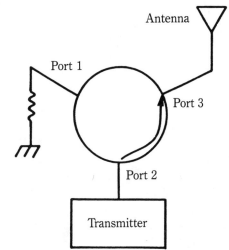

**15-4** Circulator used as an isolator.

# Study questions
# UHF and above

1. As the frequency is increased, the capacitive reactance _____.

2. In order to reduce stray capacitance, wire leads must be _____.

3. The carbon resistor will appear as a _____ and _____ at UHF.

4. _____ is a thin sheet of copper that has lower resistance at UHF than wire or cables.

5. The skin effect is the tendency of RF to exist on the _____ of the wire.

6. _____ are more efficient at microwave frequencies than are coaxial transmission lines.

7. _____ are not used at low frequencies because their size would be prohibitive.

8. The electron _____ time becomes critical as the frequency increases.

9. The _____ is a UHF and above amplifier tube. An electron beam interacts with the magnetic field of a wire.

10. The input of a microwave signal is inserted at the _____ end of the helix.

# 16
# Motors and generators

A generator converts mechanical energy into electrical energy through magnetic induction.

## The ac generator

When a conductor is moved through a magnetic field, a voltage is induced in the conductor. As the electrical conductor (*armature*) rotates through the magnetic field, magnetic lines of force are cut at an ever-changing angle. When the magnetic lines of force are cut at right angles, maximum voltage is induced in the armature. When the armature moves parallel with the lines of force, fewer lines are intersected and minimum voltage is induced. Therefore, as the armature rotates through one complete revolution within the magnetic field, a sine wave is produced. In actuality, the armature is a coil of wire. A *slip ring* is connected to each end of that coil. Because the slip rings are mounted to the shaft of the armature, they rotate with it. As the armature coil rotates through the magnetic field, a varying voltage is developed at the slip rings. This sine wave is then transferred to the load by stationary *brushes* that make constant contact with the slip rings as they rotate.

Figure 16-1 illustrates a simple generator. The field coil creates the magnetic lines of force that the armature rotates through. The two ends of the armature coil are connected to slip rings. The stationary brushes, in contact with the moving slip rings, remove the varying armature voltage. Each full 360° revolution of the armature produces one complete cycle of a sine wave, that can be monitored at the brushes.

Figure 16-2 shows the 360° circle of armature rotation clipped and stretched out into a straight line. Notice the sine wave superimposed over the line. Above the line, notice the various armature positions that produce the sine wave. In positions

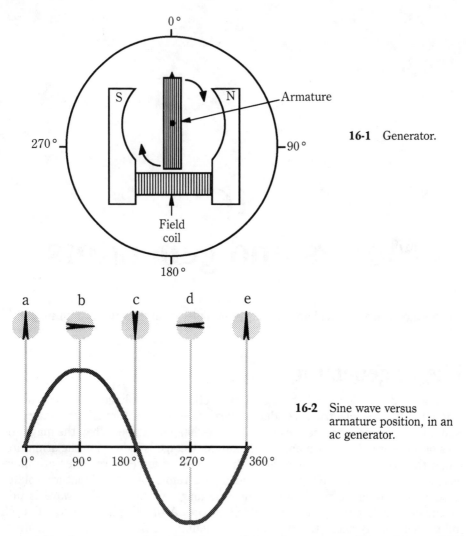

**16-1** Generator.

**16-2** Sine wave versus armature position, in an ac generator.

a, c, and e, the armature is moving in parallel to the magnetic lines of force. This produces minimum induced voltage. The maximum voltage is induced in the moving armature when it is in positions b and d. Position b offers maximum positive voltage, at 90° into the sine wave. Position d offers maximum negative voltage, at 270° into the sine wave. If the armature rotates through 60 complete revolutions in one second, the frequency of the ac sine wave will be 60 Hz.

# The dc generator

The complete sine wave, with its positive and negative polarities is present at the slip rings and brushes. This can be monitored with a voltmeter. Imagine that you could reverse the leads of the voltmeter each time the voltage reached zero and started to go negative. Figure 16-3 illustrates the waveform that would be present.

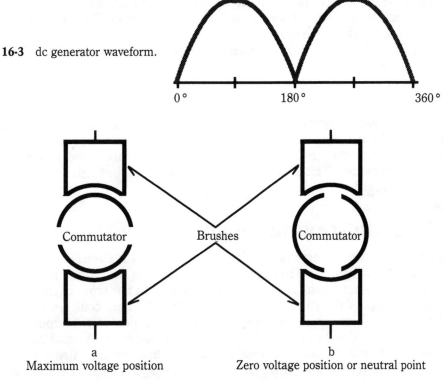

**16-3** dc generator waveform.

0° 180° 360°

a
Maximum voltage position

b
Zero voltage position or neutral point

**16-4** Commutator and brushes.

This mechanical switching is accomplished by a modified slip ring called a *commutator*. The commutator is a split slip ring that is positioned in such a manner that when the voltage from the rotating armature is at zero, the brushes switch and make contact with the opposite halves of the rings. This is the same as manually switching the leads. The commutator with the brushes acts like a mechanical rectifier, converting alternating current into pulsating direct current. Figure 16-4 shows the commutator in two positions. Figure 16-4a shows the brushes in the position of the brushes at the zero voltage point of the sine wave. This is the *neutral point*, where the mechanical switching occurs. If the positioning is not exactly at the zero voltage point, sparking might occur, because the two halves of the commutator are being shorted by the brushes. Sparking is a potential problem in generators, as it can cause:

- Radio frequency interference
- Power loss to the load
- Rapid wear of the brushes

Sparking is caused by any of the following:

- A dirty commutator
- Worn-out brushes
- Brushes in the neutral position
- Excessive play in the brushes

Sparking can be reduced or eliminated by the use of the following:
- Bypass capacitors. (These capacitors help to reduce RF interference by bypassing the RF to ground.)
- RFC. (RF chokes act as low-pass filters, passing dc but blocking radio frequencies.)
- Filter

*NOTE*: It is especially important to install an RFC between the transmitter and generator. As the frequency increases, the inductive reactance in a coil also increases. Hence, the RF choke offers a very high resistance to radio frequencies and *chokes* them out. This prevents radio frequency (RF) feedback that can cause insulation breakdown in the wires.

# The shunt-wound generator

In *separately excited* generators, the field coils receive their electrical energy from an external power source. The shunt-wound generator is *self-excited*, meaning that its field coils do not require an external power source. In this type of generator, the armature supplies the current for the field coils. The field coils are in parallel (shunt) with the armature and the load. The production of an output voltage depends upon *residual magnetism* in the field electromagnets. A small movement in the armature will cut a few residual magnetic lines of force that are always present. This results in a small induced voltage and a small current flowing through the armature. Some of that current will flow through the field coil to produce a stronger magnetic field. This stronger field causes the armature current to increase because more magnetic lines of force are being cut. The voltage gradually increases to its rated level over a period of 10 to 15 s. The shunt-wound generator is not easily affected by changing load conditions, therefore, it has good voltage regulation. The output voltage is controlled with a rheostat in series with the field coil. By increasing the resistance, the current flow in the field coil is reduced. Less current produces less magnetic field. This, in turn, causes less induced voltage in the armature.

# dc motors

While the generator converts mechanical energy into electrical energy, the motor does the opposite. It converts electrical energy into mechanical energy. When a current-carrying conductor is placed in a magnetic field, it will react with the field in such a way that it will rotate to a point that is at right angles to the field. Thus, a rotary movement is produced. By using a commutator and brushes, a continuous rotary movement is established. There are three types of dc motors: shunt motors, series motors, and compound motors. Table 16-1 summarizes their properties.

**Table 16-1. Various motor characteristics.**

| Condition | Motor type | | |
|---|---|---|---|
| | *Series* | *Shunt* | *Compound* |
| Starting torque | High | Low | High |
| Starting speed | High | Low | High |
| Running speed | Determined by the load | Remains fairly constant with load variations | Constant |
| If operated without a load | Will speed up and destroy itself | Will speed up and stabilize | Will stabilize |

# Shunt motors

In the *shunt-wound* motor, the field coil is parallel with both the armature and the dc source. This current induces magnetic fields in the armature and field coils. These opposing fields initiate rotation of the armature. At this point, the armature current is high. But, as it rotates through the field coil, a counter emf (electromotive force) is developed. This current is opposite in polarity and opposes the source current in the armature. The counter emf has a *balancing effect* that helps to maintain a constant motor speed under varying load conditions. Figure 16-5 illustrates the shunt motor as it starts. Notice the high initial armature current. In practice, this excessive current is limited by *starting resistors*. When the counter emf is minimum, the armature current is maximum. As the armature speed increases, the counter emf increases (Fig. 16-6). Under constant load conditions, the counter emf and the armature current balance, and the motor speed remains constant. When the load is increased, the motor temporarily slows, resulting in less counter emf (Fig. 16-7). Less counter emf allows more armature current to flow. More armature current causes the motor to increase in speed to the point where the two currents balance out again (Fig. 16-6). When the load is reduced (Fig. 16-8), the motor

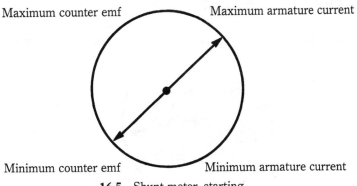

**16-5**  Shunt motor, starting.

speeds up. This causes increased counter emf and less armature current, which again stabilizes the motor speed (Fig. 16-6). This is how regulation is accomplished. Shunt motors must be protected by fuses to protect against excessive armature current. A shunt motor will actually destroy itself if not properly fused.

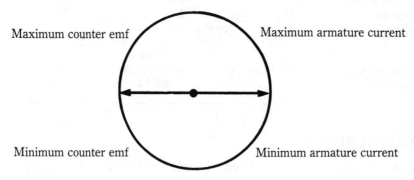

Maximum counter emf     Maximum armature current

Minimum counter emf     Minimum armature current

**16-6**   Shunt motor, running.

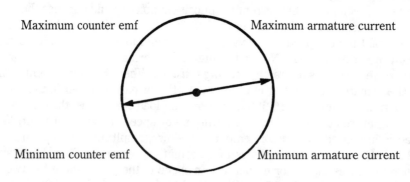

Maximum counter emf     Maximum armature current

Minimum counter emf     Minimum armature current

**16-7**   Shunt motor with increased load.

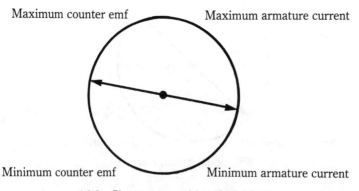

Maximum counter emf     Maximum armature current

Minimum counter emf     Minimum armature current

**16-8**   Shunt motor with reduced load.

Here is how it can happen:

The motor is running without a load; its speed will stabilize as in Fig. 16-6. If the field coil develops an open, the balancing effect of the counter emf will be lost because the counter emf will drop to nearly zero (Fig. 16-9). This allows the armature current to increase excessively. This increases the speed of the motor without limit. The motor could increase its speed to the point where it destroys itself.

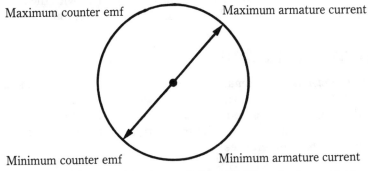

**16-9** Shunt motor with an open field coil (with no load on motor).

## Series motors

The *series-wound* motor offers higher starting speed and torque than the shuntwound motor. In this motor, the field coil is in series with the armature. Because the counter emf is less in a series motor, it does not have the same balancing effect as in the shunt motor. Hence, the speed of the series motor is controlled by its load. As the load increases, the motor slows down. As the load decreases, the motor speeds up. If the load is suddenly removed, or if the motor is operated without a load, the motor speed will increase steadily until the motor destroys itself. For this reason, gears are usually preferred over belts or chains for connecting the motor to its load. Gears are less likely to break.

## Compound dc motors

*Compound motors* incorporate both series and shunt field coils. This gives them the good regulation characteristics of the shunt motor and the high starting speed and torque characteristics of the series motor. The speed is adjusted with a rheostat in series with the shunt field coil.

## The dynamotor

The *dynamotor* is a motor and generator combination in a single unit. It has two armatures on a single shaft. Because there is only one field coil, it produces the magnetic flux for both armatures. The dynamotor is used in portable or mobile installations. It is used to step up dc voltage from a low value to a higher value.

Direct-current from a low-voltage battery is stepped up to several hundred volts. The dynamotor can be modified to produce an ac output by replacing the commutator with slip rings. There are two methods for controlling the output of a dynamotor:

- Control the dc input from the battery with a rheostat. This also changes the speed of the motor.
- Place a rheostat in series with the output.

Both methods result in some loss of power and reduction of the voltage regulation. The output is controlled by placing a rheostat in series with the dc battery at the input. A rheostat can also be placed in series with the output circuit.

# The motor-generator

The *motor-generator* is a combination of an electric motor and a generator. The motor is electrically connected to an external power source, and mechanically connected to the generator. The combination forms a power supply. It has certain advantages and disadvantages over the transformer-rectifier power supply. Table 16-2 summarizes these differences.

**Table 16-2. Motor-generator supply versus transformer-rectifier supply.**

| Motor-generator supply | Transformer-rectifier supply |
| --- | --- |
| Rugged construction | Less rugged construction |
| Operates from ac or dc source | Must operate from ac source |
| Rectifier tubes not required | Tubes require occasional replacement |
| Small filtering requirements | Requires more filtering |
| Frequent need for service | Rarely needs service |
| High initial cost | Low initial cost |
| Large and bulky | Small and compact |
| Cannot produce high voltage | Can produce high voltage |

# Study questions
# Motors and generators

1. In a dc generator, the commutator with its brushes act like a _____.

2. RF interference, power loss and rapid wear of the brushes can be caused by _____.

3. Sparking is caused by:
   a. _____
   b. _____
   c. _____

4. Bypass capacitors and RF chokes are used to eliminate _____.

5. RF feedback causes damage to _____. An RF choke is used to reduce the feedback.

6. A shunt-wound generator depends on the principle of _____.

7. When a current-carrying conductor is placed in a magnetic field, it will _____ to a position where it is at a _____ angle to the field.

8. If the field coil in a shunt-wound motor develops an open, the motor will _____.

9. The speed of a series-wound motor is controlled by _____.

10. The dynamotor can be used _____.

11. The output of a dynamotor is controlled with _____.

12. The _____ is a combination of an electric motor and a generator.

13. The motor-generator has a _____ construction but a _____ initial cost.

14. The transformer-rectifier power supply requires _____ service than the motor-generator.

15. The _____ is not capable of producing high voltage.

16. Compare the motor-generator supply with the transformer-rectifier supply.

# 17
# Measurements

This chapter entails the many methods of obtaining accurate circuit measurements.

## Measurement of voltage

Voltage is measured with a voltmeter. The voltmeter is formed by adding a series multiplier resistor to a basic meter movement. In measuring voltage in a circuit, the voltmeter (V) is connected in parallel with the load as shown in Fig. 17-1.

## Measurement of current

Current is measured with an ammeter. The ammeter is formed by adding a shunt (parallel) resistor to a basic meter movement. The ammeter (A) is connected in series with the load as shown in Fig. 17-2.

## Measurement of power

Power can be computed if a voltmeter is available and if the resistance values in the circuit are known. In Fig. 17-3, the voltmeter is placed across the resistors. Then Ohm's law is used as follows:

$$\text{Power} = \frac{E^2}{R} = \frac{120 \times 120}{110 \text{ k}\Omega} = 0.13 \text{ W}$$

Another example of calculating power with only a voltmeter is seen in Fig. 17-4. Note that in a series circuit the current flow through both resistors is the same. So,

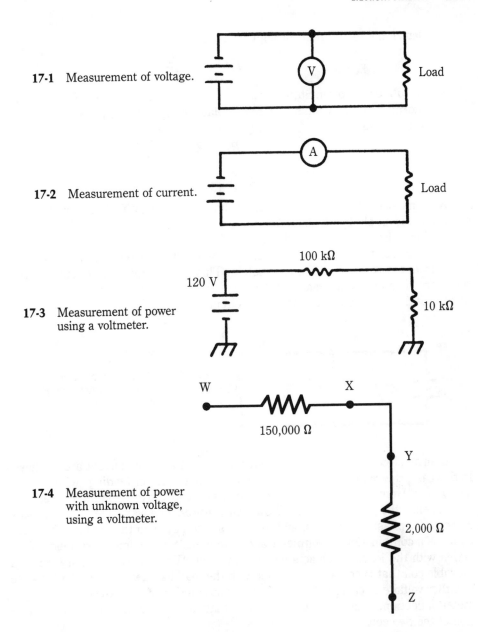

**17-1** Measurement of voltage.

**17-2** Measurement of current.

**17-3** Measurement of power using a voltmeter.

**17-4** Measurement of power with unknown voltage, using a voltmeter.

if you can determine the current flow through either resistor, you can determine the power by using Ohm's Law. Here are the steps to determine the power:

1. Decide on which resistor to measure the voltage across. It does not matter. For this example, use R1.
2. Measure voltage ($E$) across R1 with a voltmeter by measuring from point W to point X.

3. Determine the current through R1 as follows:

$$I = \frac{E}{R_1} = \frac{E}{150,000}$$

4. Calculate the power. Step 3 revealed the current flow through R1. With that information, use Ohm's Law to calculate the power dissipated by the total resistance as follows:

$$P = I^2 R$$

or

$$P = I^2 \; 152,000$$

So, what must be done is: measure the voltage from W to X, divide by 150,000, square and multiply by 152,000. You are encouraged to be very familiar with the use of Ohm's Law in situations like this.

Because power is the product of current and voltage ($P = I \times E$), it can be computed from the readings of the two meters. This is true only in dc circuits. When there are reactive components, make corrections for phase angles. Figure 17-5 illustrates the meter connections in a dc circuit.

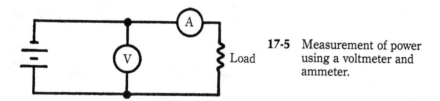

**17-5** Measurement of power using a voltmeter and ammeter.

If an ac circuit contains pure resistance, the voltage and current are in phase. In this case, the values of voltage and current from the meter readings will give you *true power*. However, if the circuit contains any capacitive or inductive reactive components, the product of *I* and *E* is *apparent power*. The wattmeter automatically corrects for phase differences, and measures true power in the circuit. The wattmeter is a combination of voltmeter and ammeter. It has a stationary field coil in series with the load, which acts like an ammeter. Within that stationary coil is a movable coil that is connected parallel with the load like a voltmeter. An increase in either voltage or current causes a greater interaction of the two coils and larger meter deflection. Phase differences are automatically accounted for by the interaction of the two coils.

# Measurement of electrical energy

Electrical energy is measured in watthours. The measuring device is called a *watthour meter*. Its operation is on the same principle as the wattmeter. However, instead of the two coils interacting to deflect an indicator needle, the coils rotate the armature of a small electric motor. The motor drives a counter that is calibrated in watthours of kilowatt hours.

# Measurement of resistance

Electrical resistance is measured by an ohmmeter. The ohmmeter is connected in series with the resistors being measured. It has a built-in dc supply that becomes a part of the circuit. The ohmmeter acts much the same as an ammeter. In fact, the meter actually measures electrical current, but its meter scale is calibrated in ohms. The ohmmeter is also used to determine if a circuit has a short or an open connection.

# Measurement of frequency

The wavemeter can be used to do any of the following:
- To check tank-circuit frequency
- To check the field strength of an antenna
- To check the output frequency of a transmitter

The wavemeter contains an LC tank circuit and an indicator meter or neon lamp. The capacitor in the LC circuit is variable. The device is placed near an oscillator, transmitter, or antenna and tuned to resonance. When the wavemeter tank circuit resonates with the circuit being measured, it absorbs power from the circuit and causes a peak in its meter reading or neon lamp intensity.

Keep the wavemeter as far away as possible from the circuit being measured. This reduces the possibility of loading the measured circuit and also increases the accuracy.

## Grid-dip meters

If the circuit you want to measure is in operation, you cannot use the wavemeter. However, if the circuit is not operating, use a grid-dip meter to measure its resonant frequency. The grid-dip meter consists of an oscillator circuit with a milliammeter connected in series with its grid-leak resistor. When an oscillator is operating, grid-leak current flows through the grid-leak resistor to ground. If the grid-dip probe is placed near an LC tank circuit with the same resonant frequency, the external LC circuit will absorb energy from the grid-dip meter and decrease the amplitude of its oscillation. This will be reflected in less grid-leak current and therefore, a dip on the milliammeter. As always, loose coupling increases the accuracy because of narrow bandwidth. (Review *Coupling Between Stages* in Chapter 9). The grid-dip meter is also used to locate undesirable parasitic oscillations in a transmitter.

## Heterodyne frequency meters

Heterodyne frequency meters contain a calibrated oscillator of known frequency, a mixer tube, and a set of earphones. The known frequency is inserted at another mixer grid. The two signals are *heterodyned* or mixed together. When any two signals are heterodyned, the sum and the difference of the two original signals will be present. It is the difference frequency that is heard in the earphones. As the two frequencies become closer, the audio signal becomes lower in frequency until a zero beat is formed when the two frequencies are equal.

## Frequency tolerance

FCC rules determine how close to the assigned frequency a transmitter must be. Frequency tolerance is stated in percentage or parts per million (ppm).

*Percentage*    For example, if a 9 MHz carrier has a frequency tolerance of 0.01%, the amount of deviation is calculated as follows:

$$0.0001 \times 9 \times 10^6 \text{ Hz} = 900 \text{ Hz}$$

In other words, the 9 MHz carrier would have to stay within 900 Hz of its assigned frequency to be within tolerance as follows:

$$9,000,000 \text{ Hz} + 900 \text{ Hz} = 9,000,900 \text{ Hz}$$
$$9,000,000 \text{ Hz} - 900 \text{ Hz} = 8,999,100 \text{ Hz}$$

*Parts per million (ppm)*    To determine the ppm, move the decimal point six places to the right. Therefore, 0.0001 is equal to 100 ppm. To determine the tolerance, multiply the number of MHz by the ppm as follows:

$$9 \text{ MHz} \times 100 = 900 \text{ Hz}$$

Often the RF oscillator is followed by a number of frequency doublers or triplers to achieve the desired output frequency of the transmitter. To determine the frequency tolerance of the oscillator, divide the frequency tolerance of the output carrier frequency by the number of multiplier stages. For example, if the oscillator operates at one eighth (three doublers would multiply the oscillator $2 \times 2 \times 2 = 8$ times) of the output frequency, divide the tolerance by 8, as $900/8 = 112.5$ Hz. In this case, the oscillator will only deviate 112.5 Hz. If the oscillator operates at one sixth of the output frequency, divide by 6, as $900/6 = 150$ Hz. The oscillator cannot drift more than 150 Hz or the output will drift more than 900 Hz. Figure 17-6 shows a transmitter with a 2182 kHz output frequency. The transmitter has three doublers, making the oscillator one eighth of the output frequency (272.75 kHz). The FCC Rules and Regulations 23.16 states that the frequency tolerance must be 15 ppm. As stated above, you multiply the ppm by the frequency in MHz. In this case, you must convert 2182 kHz to MHz, by moving the decimal point three places to the left, giving you 2.182 MHz. The output can drift 15 ppm $\times 2.182 = 32.73$ Hz. The allowable drift of the oscillator must not exceed $1/8 \times 32.73 = 4.09$ Hz. If you multiplied 0.27275 MHz times 15 ppm you would arrive at the same answer.

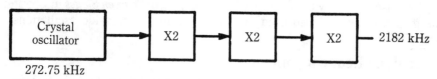

272.75 kHz

**17-6**    Crystal oscillator operating at one-eighth of output frequency.

Another example might prove helpful. The transmitter has an assigned output frequency of 12 MHz. The allowed tolerance is 100 ppm. If the oscillator operates at one fourth of the output frequency, how much can the oscillator drift?

12 MHz × 100 (ppm) = 1200 Hz maximum allowable deviation of the output. 1200/4 = 300 Hz Maximum allowable drift in the oscillator. This 300 Hz drift will be multiplied four times by the frequency multiplier stages, resulting in an output frequency drift of 1200 Hz.

# Calculation of decibels

The decibel is a logarithmic ratio of one power, voltage, or current to another.

*Power* When comparing two power levels, the following formula is used:

$$dB = 10 \log \frac{P}{P}$$

It does not matter which power level is divided by the other because the answer will be the same, with the only exception being that one value will be a negative value. It will be quite obvious by the starting and ending power level, whether the change is a power gain or attenuation. A 3-dB gain equates to doubling the power. Passing a signal through an attenuator that is rated − 3 dB results in a 50% reduction in the power of that sign.

Refer to the 10 - 20 - 30 method and to the other dB tips in Chapter 14. That chapter deals extensively with power changes, that relate to effective radiated power. The following examples will help you understand how to calculate other types of dB problems.

*Example* An amplifier has an input of 5 W and output of 50 W. How much gain, in decibels, is present in the amplifier?

$$dB = 10 \times \log \frac{50 \text{ W}}{5 \text{ W}}$$
$$= 10 \times \text{the log of 10 (the log of 10 is 1).}$$

Therefore 10 × 1 = 10 dB gain.

Another way to calculate this is to divide 50 W by 5 W. You arrive at a power change of 10x, which equates to a 10 dB gain.

*Example* With single sideband suppressed carrier (J3E) transmission, the carrier must be suppressed 40 dB below peak envelope power. If the output power of the transmitter is 100 watts, how much power may be in the carrier?

As discussed, a − 40 dB attenuation equates to a power change of 1/10,000 x (or 0.0001 x). Hence, 100 W × 0.0001 = 0.01 W. The carrier must be suppressed to .01 watts. This can be verified by the following formula:

$$dB = 10 \log \frac{100 \text{ W}}{0.01 \text{ W}}$$
$$= 10 \times \text{the log of 10,000}$$
$$= 10 \times 4 = -4 \text{ dB}$$

*Voltage* When comparing two voltage levels, use the following formula:

$$dB = 20 \log \frac{E}{E}$$

*Example*   If you start with a 1200-μV signal and attenuate it to 120 mV, what is the decibel change (loss)?

$$dB = 20 \times \log \frac{1200 \times 10^{-6}}{120 \times 10^{-3}}$$
$$= -40 \text{ dB, or a 40 dB loss}$$

*Current*   When comparing two current levels, use the following formula

$$dB = 20 \log \frac{I}{I}$$

*NOTE:* If you get a decibel problem on the FCC test, plug the multiple-choice answers into the various formulas shown above. Beware of the difference between power calculations and voltage or current calculations (10 log versus 20 log). In other words, verify your answer.

# The oscilloscope

The oscilloscope is probably the most useful and versatile piece of test equipment available. With it, you can measure voltage, frequency, phase differences, and much more. What makes the oscilloscope so unique is that the actual waveforms are graphically displayed on a CRT (cathode ray tube) display. By observing the waveform, you can see noise or distortions that other test equipment would miss.

On the face of the CRT display, is a *graticule*, a grid that is divided into vertical and horizontal divisions. The two most commonly used controls on the scope are the vertical sensitivity and the horizontal sweep selector. The vertical control adjusts the sensitivity in the vertical direction and is used for measuring voltage. The horizontal control is used for the measurement of time, which may be easily converted to frequency.

Two commonly used functions of the oscilloscope are the measuring of voltage and the measuring of frequency. The methods of doing so are described as follows:

**Measuring dc voltage**   When the scope is turned on, with no voltage applied to the vertical input, a small spot appears on the center of the grid. When a positive dc voltage is applied to the vertical input, the spot moves vertically upward. When a negative dc voltage is applied, the spot moves vertically downward on the grid. The voltage is determined by multiplying the number of divisions the spot has moved by the volts/division setting on the vertical sensitivity control.

For example, if the spot moves up two full divisions and the sensitivity is set for 5 volts/division, the voltage is +10 V. Figure 17-7 illustrates a spot in the center, representing V and a spot up two divisions, representing +10 V, in this example.

**Measuring ac voltage**   When an ac voltage is applied to the vertical input, the spot moves up and down on the display. This creates a solid vertical line, because of the rapidity of the polarity change. For example, if household voltage is measured, the spot moves from the full positive position to the full negative position and back 60 times each second, because the line frequency is 60 Hz. The voltage measured is called *peak-to-peak* voltage. But the full potential of the scope is realized when the actual ac waveform is viewed on the display. When the waveform is displayed, its frequency can be measured.

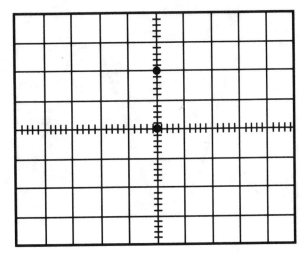

**17-7**  Measuring dc voltage with an oscilloscope.

**Measuring frequency**  By adjusting the horizontal sweep selector, the spot can be made to move rapidly from left to right, forming a solid line. This solid image is called the *trace*. The period of time required for the spot to sweep across the screen is controlled by the horizontal sweep selector, which is calibrated in seconds/division.

Assume a sine wave occupies eight divisions with the horizontal sweep selector set to 2 milliseconds per division. By multiplying the horizontal sweep setting by the number of divisions, you have 2 milliseconds times 8 divisions, or 16 milliseconds. Once the time required for one complete sine wave cycle is known, the frequency of the sine wave can be calculated as follows:

$$\text{Frequency} = \frac{1}{\text{time}}$$
$$= \frac{1}{16 \text{ ms}}$$
$$= \frac{1}{0.016 \text{ sec}}$$
$$= 62.5 \text{ Hz}$$

Figure 17-8 illustrates a sine wave that takes six divisions to complete a cycle. With the horizontal sweep set at 50 microseconds per division, the frequency is calculated as follows:

$$\text{Frequency} = \frac{1}{\text{time}}$$
$$= \frac{1}{0.0003}$$
$$= 3333.3 \text{ Hz or } 3.33 \text{ kHz}$$

The following is a brief summary of oscilloscope components. Because the CRT is the heart of the instrument, the various components are described in reference to the CRT.

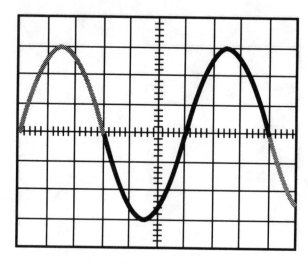

**17-8** A 3.33 kHz sine wave. Horizontal sweep is set to 50 microseconds per division.

**Cathode**   Produces electrons.

**Grid**   Functions much like the grid of a triode tube. The amount of electrons that pass the grid is influenced by the potential on the grid. Therefore, the intensity of the beam is controlled by the grid.

**Focusing anode**   Controls the focus of the electron beam.

**Accelerating anode**   A highly positive cylinder that attracts the electrons and accelerates them toward the CRT screen.

**Deflection plates**   Vertical and horizontal deflection plates deflect the electron beam in accordance with the amplitude of the voltage applied to them.

The vertical deflection plates are driven by a vertical amplifier. Increased voltage applied to these plates causes the trace to move vertically, as an indication of amplitude of the input signal. The horizontal plates are driven by a sweep generator, which causes the trace to move horizontally, enabling the technician to measure frequency of the input signal.

**Aquadag coating**   A conductive coating on the inside of the CRT that extends to the screen. It is connected to the highly positive accelerating anode and therefore has the following effects:

• Shields the electron beam from external electric fields
• Produces a final acceleration of the electron beam
• Attracts the secondary emission electrons from the screen

**Fluorescent screen**   A phosphorescent coating is applied to the inside of the screen. When the electron beam strikes this coating, a bright spot is formed. As the bright spot is moved by the input signal, a waveform is produced. The bright spot must not be left in one spot for a long period of time or a localized burn can result on the fluorescent coating, rendering the screen unresponsive in that area.

# The spectrum analyzer

The spectrum analyzer or spectrum monitor is similar to the oscilloscope. The oscilloscope presents an amplitude versus time display, and the spectrum analyzer presents an amplitude versus frequency display. This characteristic enables you to determine the frequency of the radio signal directly off the CRT display, without calculations. The horizontal sweep is calibrated in terms of frequency per division, instead of time per division.

The CRT display allows you to view the radio signal, with its associated harmonics, parasitics or other spurious signals. The heights of the images reveal the relative amplitudes.

If you viewed a 10-kHz sine wave, with no spurious signals, on a spectrum analyzer, you would see one spike at the 10 kHz location. If you viewed a 10-kHz square wave, you would see an image at 10 kHz, 30 kHz, 50 kHz, 70 kHz and other odd harmonics, because square waves comprise numerous odd harmonics. If you viewed a 10-kHz triangular wave, you would see numerous even harmonics, like 20 kHz and 40 kHz, on the display.

# Measurement of temperature

Society is in transition from the decimal system to the metric system. Proper operation of equipment often depends on the temperature of the environment. In avionics or marine work, it is important to know how to quickly convert between Celsius and Fahrenheit temperatures.

A quick formula for converting from Fahrenheit (F) to Celsius (C) is:

$$\text{Celsius} = (F - 32) \times 0.555$$

For example, to convert 50° F to Celsius, do the following:

$$\text{Celsius} = (50 - 32) \times 0.555$$
$$= 10\,°C$$

To convert 104 °F to Celsius, do the following:

$$\text{Celsius} = (104 - 32) \times 0.555$$
$$= 40\,°C$$

A quick formula for converting from Celsius to Fahrenheit is:

$$\text{Fahrenheit} = (C \times 1.8) + 32$$

For example, to convert 40 °C to Fahrenheit, do the following:

$$\text{Fahrenheit} = (40 \times 1.8) + 32 = 104\,°F$$

A 100 °C temperature is converted to Fahrenheit as follows:

$$\text{Fahrenheit} = (100 \times 1.8) + 32 = 212\,°F$$

Here is a tip that may make it easier to remember this. Remember that the freezing point of water is 32 °F and 0 °C. The Celsius temperatures are always smaller than their Fahrenheit counterparts. When converting from Fahrenheit to Celsius, you must make the number smaller in two ways—by subtracting 32 from the number and by multiplying by a number less than one (0.555).

When converting from Celsius to Fahrenheit, you must make the number larger—by multiplying by 1.8 and by adding 32. If you forget the 1.8, just punch 0.555 on your calculator and press the 1/X key.

# Study questions
# Measurements

1. What formula is used when measuring power in a circuit where the values of resistance are known and only a voltmeter is available? _____.

2. How are a voltmeter and an ammeter connected to a circuit to measure power? _____

3. A wattmeter measures _____ power.

4. In the wattmeter, power is the product of _____ and _____. Automatic correction is made for _____.

5. Electrical energy is measured by a _____.

6. Electrical resistance can be measured by _____.

7. What device is used to determine whether a short or open circuit exits? _____.

8. What three things can an absorption wavemeter measure?
   a. _____
   b. _____
   c. _____

9. What precaution should be taken into account while using a wavemeter? _____

10. The grid-dip meter can measure the _____ of an LC tank circuit.

11. The output frequency of a transmitter is 156.8 MHz and the oscillator is one eighth the output frequency. If the allowable drift is 10 ppm, how much can the oscillator drift and still be within tolerance? _____

12. If the power becomes 10 times the original power, how much dB gain has taken place? _____

13. What controls the intensity of the CRT beam? _____

14. What are three uses of the aquadag coating in a CRT?
   a. _____
   b. _____
   c. _____

# 18
# Digital circuitry

Digital logic circuits respond to discrete events that fall into two categories:

| | | |
|---|---|---|
| OFF | or | ON |
| 0 | or | 1 |
| LOW | or | HIGH |

The various gates have two or more inputs and a single output. Their inputs sense either an ON or OFF condition. Their output develops either a high (ON) or a low (OFF) condition. Several commonly used gates are discussed and summarized with truth tables, followed by discussion of some circuit arrangements.

## The AND gate

To obtain a high output with an AND gate, all of its inputs must be HIGH. Any other condition will result in a low output. Table 18-1 summarizes all the possible combinations. Figure 18-1 illustrates the circuit symbol for the AND gate.

## The NAND gate

The NAND gate produces a high output with every combination except when both inputs are high. It is opposite to the AND gate. Table 18-2 summarizes the possibilities. Figure 18-2 illustrates the circuit symbol for the NAND gate.

## The OR gate

To obtain a high output with an OR gate, only one of its inputs need to be high. Table 18-3 summarizes the possibilities. Figure 18-3 shows the circuit symbol.

**Table 18-1. Truth table for AND gates.**

| Input 1 | low | high | low | high |
|---------|-----|------|-----|------|
| Input 1 | low | low | high | high |
| Output | low | low | low | high |

Input
1
2
Output   **18-1**   AND gate.

**Table 18-2. Truth table for NAND gates.**

| Input 1 | low | low | high | high |
|---------|-----|-----|------|------|
| Input 2 | low | high | low | high |
| Output | high | high | high | low |

Input
1
2
Output   **18-2**   NAND gate.

**Table 18-3. Truth table for OR gates.**

| Input 1 | low | high | low | high |
|---------|-----|------|-----|------|
| Input 2 | low | low | high | high |
| Output | low | high | high | high |

Input
1
2
Output   **18-3**   OR gate.

# The NOR gate

To obtain a high output from a NOR gate, both inputs must be low. In the OR gate, any high input would produce a high output. The NOR gate is the opposite. Any high at the input will produce a low at the output. And a high is produced with no input at all. Refer to Table 18-4 for the NOR gate truth table. Figure 18-4 shows the circuit symbol.

**Table 18-4. Truth table for NOR gates.**

| Input 1 | low | low | high | high |
|---------|-----|-----|------|------|
| Input 2 | low | high | low | high |
| Output | high | low | low | low |

Input $\begin{array}{c} 1 \\ 2 \end{array}$ ———— Output    **18-4**   NOR gate.

# The NOT gate

The NOT gate acts like an inverter. If a high appears at its input, a low will appear at its output. A low input will produce a high output. Table 18-5 truth table summarizes all the possibilities. Figure 18-5 shows the NOT gate.

Figure 18-6 offers four combinations of gates for practice.

**Table 18-5.**
**Truth table for NOT gates.**

| Input | low | high |
|---|---|---|
| Output | high | low |

————— **18-5**  NOT gate.

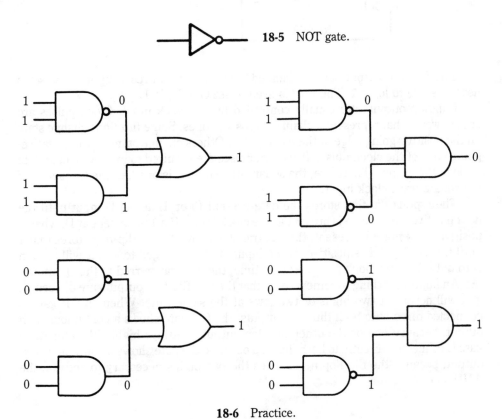

**18-6**  Practice.

# Flip-flops

The flip-flop is a digital device that has numerous applications. They may be used for information storage and in devices such as binary counters. The flip-flop is capable of recognizing zero (a low) or one (a high). The outputs are also measured in terms of zero and one.

Figure 18-7 illustrates a type-D flip-flop. The *D* stands for delay, or data, because it stores and delays one bit of data at a time. The Q and NOT-Q are complementary outputs, meaning that when one output is 0, the other is 1, and vice versa. The clock input may be abbreviated as CL, CK, T, or C.

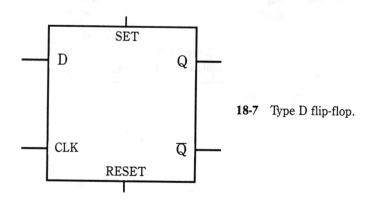

**18-7**   Type D flip-flop.

The flip-flop output can be changed with the set and reset inputs. The *set* is used to set Q to logic 1. The *reset* is used to set Q to logic 1.

Flip-flop outputs are usually controlled by the clock input. Clock pulses are square waves that represent a chain of zeros and ones. Some flip-flops change state on the positive-going edge of the clock pulse. Others change state on the negative-going edge of the clock pulse. Two clock pulses are required to produce a complete cycle at the output. Therefore, the square wave output is one half the frequency of the square wave clock input.

The type-D flip-flop stores the data present (0 or 1) at the D input until the next positive clock pulse is present at the clock input. If a 1 is present at D, when a positive clock pulse is present, the 1 is transferred to Q. The flip-flop stores that 1 until the clock pulse is positive. If the input at D is changed to a 0, it will remain zero until the next clock pulse. At that time, the 0 is transferred to the Q output.

An important thing to remember is that the two flip-flop outputs are opposites. You will never get two highs or two lows at the same time. When an OR gate is connected directly to both flip-flop outputs, the OR gate shows a continuous high output, because one or the other flip-flop output is always high. When an AND gate is connected directly to both flip-flop outputs, the gate shows a continuous low output, because the flip-flop never gives the two high's necessary to activate the AND gate.

# Binary numbers

The language of digital circuits consists of 0s and 1s—highs and lows. Decimal numbers can be described in these terms as well. When done so, they are called *binary* numbers. *Binary numbers* consist of various locations that have decimal equivalents. The location value increases from right to left, with each value doubling. The values for the first four locations are 8 4 2 1. If the location is occupied with a 0, it has no decimal value. If the location is occupied with a 1, the place value, of that location, is added to any other locations that are occupied by a 1. Thus 0001 = 1; 0010 = 2; 1101 = 13; 1111 = 15; etc. Table 18-6 shows the progression of decimal and binary equivalents.

**Table 18-6. Decimal-binary equivalents.**

| Decimal | Binary | Decimal | Binary |
|---------|--------|---------|--------|
| 0 | 0000 | 8 | 1000 |
| 1 | 0001 | 9 | 1001 |
| 2 | 0010 | 10 | 1010 |
| 3 | 0011 | 11 | 1011 |
| 4 | 0100 | 12 | 1100 |
| 5 | 0101 | 13 | 1101 |
| 6 | 0110 | 14 | 1110 |
| 7 | 0111 | 15 | 1111 |

# Study questions
# Digital circuitry

Complete the missing information in the circuits of Fig. 18-8.

1. All inputs are low (in Fig. 18-9), making the output high. A high on which input will make the output low?_____

2. All inputs are low (in Fig. 18-10), making the output low. A high on which input will make the output high? _____

3. All inputs are low (in Fig. 18-11), making the output low. A high on which input will make the output high?_____.

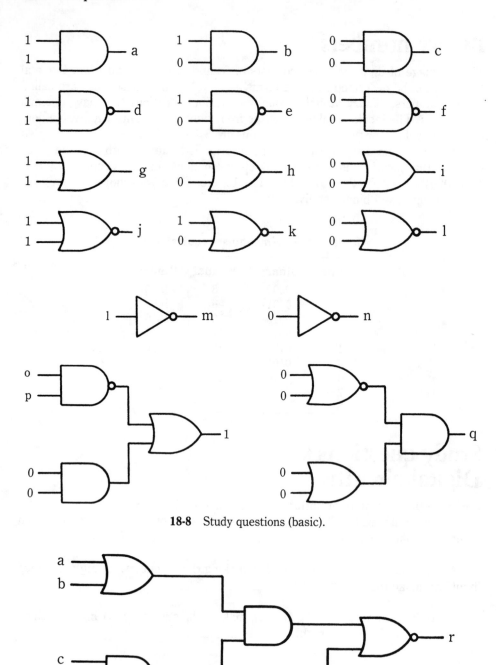

**18-8** Study questions (basic).

**18-9** Advanced study question 1.

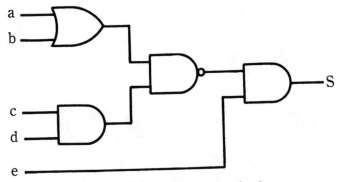

**18-10** Advanced study question 2.

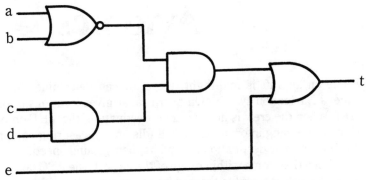

**18-11** Advanced study question 3.

# 19
# Avionics

While an aircraft is mid-flight, the crew can determine their position with the Omega Navigation system. On-board radar gives locations of other aircraft in the area. When the crew is nearing their destination, the VHF omni-range (VOR) provides positioning information. DME (distance measuring equipment) provides the distance to their destination as well as their ground speed.

When they come within range of the airport, the ATC (air traffic control) sends an interrogation signal to their on-board transponder to determine important information about the aircraft.

As the crew comes in for their final approach, the ILS (instrument landing system) provides vital positioning information. Three marker beacons provide additional information as they make their final approach to the runway.

## Omega Navigation System

The Omega Navigation System is used for worldwide navigation. It can be used by aircraft or ships. There are eight Omega transmitter stations around the world. The transmitting frequency is within the VLF (very low frequency) band, which ranges from 3 – 30 kHz. Specifically, they transmit on 9 – 14 kHz. The transmitters are precisely timed so that the phase of all eight transmitter signals is the same. On-board receiving equipment compares the phase shifts of the various signals to determine location.

## Radar

The last chapter of this book goes into considerable detail to prepare you for the Radar Endorsement. This brief coverage of radar is sufficient for the General Radiotelephone exam.

Radar stands for radio detecting and ranging. A radar transmitter sends out pulses that are reflected back by aircraft and other objects. Radio waves travel 186,000 miles per second, or 162,000 nautical miles per second. Radio waves travel 328 yards per microsecond. Because the radar pulse must travel round trip to the object and back to the radar receiver, the true distance of the object is 164 yards per microsecond. The radar receiver converts the microsecond time delay of the echo signal into distance that is visible on a CRT display. Each one microsecond equates to a target distance of 164 yards. Because there are 2027 yards in a nautical mile, a target one nautical mile away has a delay of 12.4 microseconds (2027/164 = 12.4). This is called a *radar mile*. This method of determining the distance of the target is called *range determination*.

The method of determining the direction of the target is called *azimuth* or *bearing determination*. The azimuth of the target, or horizontal position on the 360° horizon, is in reference to the ship's heading or to true north. When the reference is true north, it is called *true bearing*. When the reference is the ship's heading, it is called *relative bearing*.

# VOR (VHF omni-range)

VOR transmitter stations are placed at various ground locations. The VOR receiver is a component of the aircraft navigation system. Together, the VOR transmitter and VOR receiver give valuable information pertaining to the position of the aircraft, in relation to the VOR transmitter site.

The *V* stands for VHF, and gives away the general frequency range of 30 – 300 MHz. More specifically, in the VHF band, VOR transmitters are assigned operating frequencies between 108 MHz and 118 MHz.

The VOR transmitter frequency modulates a subcarrier with a 30-Hz tone. That signal is sent to a directional antenna that is electronically rotated at a rate of 30 revolutions per second (1800 revolutions per minute).

The approaching aircraft VOR receiver receives an FM signal that is changing in amplitude 30 times per second, because of the rapidly rotating antenna. When the antenna is pointing at the aircraft, the signal amplitude is maximum. When the antenna is pointed in the opposite direction the amplitude is minimum.

The VOR receiver demodulates the AM and the FM signals to produce two 30-Hz audio signals that are fed into a phase detector circuit. The phase difference between the two signals tells the pilot where the aircraft is located with respect to the VOR transmitter.

The transmitter is synchronized with the rotating antenna in such a way that the phase of the FM signal is maximum positive when the antenna is pointing north, and maximum negative when the antenna is pointing south.

If the aircraft is directly north of the VOR transmitter, the FM signal, entering the VOR receiver, will be maximum positive, and the antenna-generated AM signal will be maximum positive because the antenna is pointing at the aircraft. The 30-Hz audio tones, from the demodulated FM signal and the demodulated 30-Hz AM signal would be in phase at the VOR receiver.

Because the two 30-Hz signals are in phase when the plane is north of the VOR site, it can be visualized how the phase relationship would change as the aircraft flies around the VOR transmitter. Figure 19-1 summarizes the aircraft positions with relation to the VOR site, and Fig. 19-2 illustrates the phase differences. If the antenna (AM) signal peaks 90° after the FM signal, the aircraft is east of the transmitter, as shown in Fig. 19-2b. If the antenna signal peaks 180° after the FM signal, the aircraft is south of the transmitter, as shown in Fig. 19-2c. Figure 19-2d illustrates a delay of 270°, with the aircraft west of the VOR transmitter. Remember that the 30-Hz FM signal always peaks when the antenna is pointing north, as shown in Fig. 19-2a. The 30-Hz AM signal always peaks when the antenna is pointing directly at the aircraft. Because of this, there will always be a phase difference between the two signals, unless the aircraft is directly north of the VOR site.

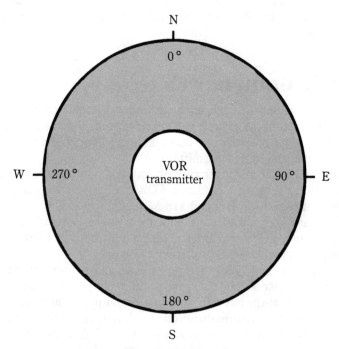

**19-1**   VOR (VHF omni-range) operation.

# DME (distance measuring equipment)

Distance measuring equipment is similar to radar in that the aircraft crew can use it to determine distance. With radar, a transmitted radar signal is reflected from an object. The time delay is displayed as a distance on the cathode ray tube. With DME, a signal is sent from the aircraft and received by a DME ground station. The ground station delays the signal, a predetermined amount, and retransmits it. The aircraft DME equipment receives it and calculates the time delay. The DME panel

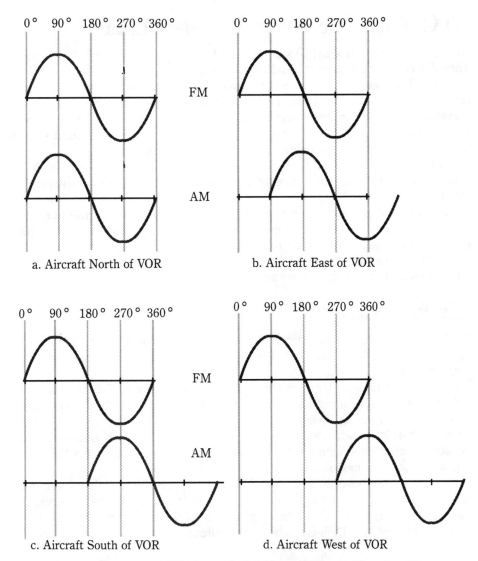

a. Aircraft North of VOR

b. Aircraft East of VOR

c. Aircraft South of VOR

d. Aircraft West of VOR

**19-2**  VOR phase relationships at VOR receiver.

presents a digital readout of the distance to the DME ground station, in nautical miles. Because the distance is constantly changing, as the aircraft approaches the DME station, the rate of change is converted into ground speed and *time-to-DME-station* indications, that are also displayed on the DME panel.

Each airborne DME transmitter sends out pulse pairs that are randomly spaced. The DME receiver, tuned to a frequency above or below the transmitter frequency, locks on its own unique signal that has been retransmitted by the ground station. Because each aircraft has its own unique signal, the DME ground station can accommodate numerous planes simultaneously.

# ATC (air-traffic control) transponders

At the air traffic control (ATC) station, two types of radar are used. The Primary Surveillance radar operates by sending radar pulses and receiving the echoes from the aircraft. No aircraft response is necessary. The secondary surveillance radar, sometimes called the ATC radar, requires a response from the aircraft. The ATC transmitter sends an interrogation pulse to the aircraft. The airborne ATC transponder transmits the aircraft identification and altitude to the ground station. Some of the transponder information is visible on the radarscope.

Air-traffic control ground stations send interrogation pulses on 1030 MHz. The ATC transponders send the reply pulses on 1090 MHz. The effective range is about 200 mi.

Although this system serves no navigational purpose for the aircraft, it enables the air traffic control personnel to more effectively manage the airspace. ATC transponders are required on all aircraft flying under instrument flight rules (IFR). Other planes are not required to have transponders, although large airports may prohibit the landing of aircraft without transponders.

# Marker beacons

Marker beacons are very directional fan-shaped signals that are directed in a vertical direction. When the aircraft flies over the beacon, colored indicator lights notify the pilot.

The beacon transmitter operates on a carrier frequency of 75 MHz. The three individual marker beacons modulate the 75 MHz carrier with different audio tones. The outer marker, which is located about 5 mi. from the runway, is modulated with a 400-Hz audio tone. The middle marker, which is 3500 ft from the landing strip, is modulated with a 1300-Hz audio tone. The inner marker, which is 100 ft from the strip, is modulated with a 3000-Hz audio tone.

In the marker beacon receiver, the 75-MHz signal is demodulated, removing the audio tones. Three audio filters are placed at the output of the receiver. The 400-Hz filter output activates a blue light. The 1300-Hz filter output activates an amber light, and the 1300-Hz audio filter output activates a white light on the instrument panel.

The outer marker marks the point where the aircraft begins its descent, with the help of the instrument landing system.

# ILS (instrument landing system)

The instrument landing system is designed to guide the aircraft to a successful landing. Visualize a steadily rising imaginary line extending from the runway to the outer marker, about five miles away. This line is to be followed by the aircraft during the approach. If the aircraft deviates horizontally, to the left or the right of the imaginary line, the localizer electronics informs the pilot. If the aircraft deviates in a vertical direction, above or below the line, the glideslope instruments inform the pilot.

The localizer ground equipment directs two narrow radio beams along the imaginary line we discussed. One signal is modulated with a 90-Hz tone. The other is modulated with a 150-Hz tone. Both signals are of equal signal strength. The 90-Hz tone is directed slightly to the left of the centerline (from the pilot's vantage point). The 150-Hz tone is directed slightly to the right of the centerline. The localizer receiver balances one signal against the other and produces a null when the aircraft is directly in the center of the two beams. When the aircraft moves to the right of center, the 150-Hz signal is stronger. When the aircraft moves to the left of the centerline, the 90-Hz signal is stronger. Visual instrumentation shows the pilot positioning in relation to the centerline.

The glideslope ground equipment uses the same 90-Hz and 150-Hz signals on a higher carrier frequency. With the glideslope, the signals are directed in such a manner that when the aircraft is above the centerline, the 90-Hz signal is stronger. When the aircraft is below the centerline, the 150-Hz signal is stronger. Visual instrumentation shows the position of the aircraft with reference to the desired center position.

The glideslope and the localizer radio beams together form an ever-narrowing rectangular tunnel for the aircraft to fly through during the approach. This is sometimes called the glideslope-localizer envelope. See Fig. 19-3.

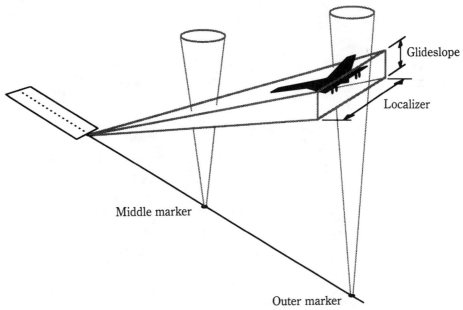

**19-3**  Glideslope localizer envelope.

# Study questions
# Avionics

1. On what frequency range does the Omega Navigation System operate?

2. How many yards does a radar pulse travel in 1 $\mu$s?

3. On what frequency band does the VOR operate?

4. What function does the glideslope instrumentation serve?

5. What does the localizer do?

# 20

# Marine radar fundamentals (FCC element 8)

Radar (radio detection and ranging) is an electronic navigational guide. It enables safe navigation in darkness, fog, or storm. The radar transmitter sends microwave pulses through a highly directional antenna system. When these radio waves encounter objects such as other ships, planes, or land masses, a portion of the signal returns to the radar unit as a reflection or echo. By evaluating the delay time of the reflected pulse, the distance of an object can be determined. The direction of the object is the same as the direction of the antenna system.

## Radar propagation

The radar waves always travel at the speed of light, which is 186,000 statute mi/s. However, the nautical mile is often used when talking about distance. A nautical mile is slightly longer than a statute mile. It is the nautical mile that is spoken of when you refer to the *radar mile*. A radar mile is the time required for the radar wave to complete a round trip (from the transmitter to the target and back). Because radio waves propagate one nautical mile in 6.2 $\mu$s, a radar mile is 6.2 times 2 or 12.4 $\mu$s for the round trip. Table 20-1 summarizes the differences between nautical miles and statute miles.

### Table 20-1. Propagation summary.

| Propagation component | Statute miles | Nautical miles |
|---|---|---|
| Number of feet | 5,280 ft | 6,080 ft |
| Miles per second | 186,000 mi/s | 162,000 mi/s |
| Distance per microsecond | 0.186 mi | 0.162 mi |

*NOTE*: Remember that radar ranging is a two-way process. That is, the radar signal must travel to the target and back. Therefore, the actual radar range velocity is one half of 0.162 miles per microsecond, or 0.081 nautical miles per microsecond. For example, if the radar unit indicates that the round-trip delay time is 123.44 $\mu$s, the actual range of the target is 123.44 times 0.081, or about 10 nautical miles.

# Radar receiver

Figure 20-1 is a simplified block diagram of a radar receiver. Study the relationships of the individual components as well as their operation. Details can be found in the glossary.

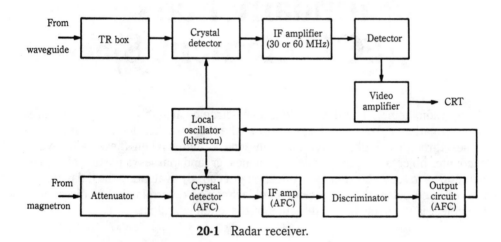

**20-1**  Radar receiver.

# Radar glossary

This glossary contains details on the complete radar system. Refer to the diagrams when necessary. It will be important to have a clear understanding of the following terms. Study them well.

**antenna**  A very narrow antenna radiation pattern is required in order to provide accurate bearing measurements. A parabolic reflector is used to focus the microwave energy in much the same way a flashlight reflector focuses the light into a narrow beam. The radiation pattern from a radar antenna has two characteristics. First, it is extremely narrow in the horizontal plane (azimuth). This allows for good bearing resolution. Secondly, the beam is wide in the vertical plane. This characteristic allows the ship to roll from side to side in the waves without losing the target.

**antenna maintenance**  The reflector should be kept clean. A thin layer of soot or dirt will have little or no effect on the performance of the ship radar. However, if a thick layer of corrosion or crust is allowed to form, the performance will be decreased, especially for weak targets.

**aquadag coating** The aquadag is a conductive coating inside the cathode ray tube. It acts as a second anode and therefore gives a final acceleration to the electron beam. It also protects the electron beam from stray electric fields. It provides no protection from magnetic fields.

**artificial transmission line** This circuit determines the duration and shape of the transmitter pulses. For good frequency response, the pulses applied to the magnetron must be a high-quality square wave.

**attenuator** Because of the high transmitter power, an RF attenuator must be placed between the magnetron and the AFC section of the receiver. This prevents damage of the AFC crystal detector.

**ATR box** The anti-transmit-receive cavity prevents the received echo pulses from entering the transmitter and being absorbed. It, like the TR box, has a spark-gap tube that ionizes when the transmitter pulses are present. Because this switch does not protect the receiver, the dc keep-alive voltage is not required because it is in the TR box.

**authorized radar frequency bands** The following are three frequency bands used for ship radar transmitters:

$$2900 - 3100 \text{ MHz}$$
$$5460 - 5640 \text{ MHz}$$
$$9300 - 9500 \text{ MHz}$$

**automatic frequency control (AFC)** The magnetron tube might drift in frequency. It is the purpose of the AFC circuits to cause the *klystron* (local oscillator) to track any changes in the magnetron frequency so that the difference frequency between the klystron and the magnetron will always be equal to the intermediate frequency.

**bearing resolution** The ability to separate two adjacent targets of equal distance. A very narrow antenna radiation pattern in the horizontal plane is required to achieve this.

**blooming** Blooming is excessively expanded blips on the PPI screen. It can be caused by excessive voltage from the IF section.

**discriminator** The discriminator is tuned to the IF (30 MHz). When the magnetron and klystron are operating properly, their frequencies are 30 MHz apart. When this 30-MHz IF is applied to the discriminator no output is produced. However, if the magnetron or the klystron should drift in frequency, the frequency applied to the discriminator will be higher or lower than the 30 MHz IF, and a voltage output is produced. This error voltage is applied to the repeller (reflector) plate of the kylstron to change its frequency in order to maintain a constant 30-MHz IF.

**Doppler effect** When an object is approaching the radar installation, the echo signal will be compressed, making its frequency slightly higher. When the object is moving away from the radar installation, the frequency of the echo signal will be slightly lower in frequency. This is called the Doppler effect. In

Doppler radar, a discriminator is used as a second detector in its receiver. As a result, only moving objects will show up on the PPI screen. Stationary objects such as trees, buildings, bridges, etc., will not be visible on the screen because they will reflect the radar pulses back at exactly the same frequency. The discriminator only responds to echo pulses that are above or below the transmitted frequency.

**duplexer** The duplexer is a switch that enables the use of a common antenna for the receiver and transmitter. It disconnects the receiver when the transmitter is sending pulses, and disconnects the transmitter when the receiver is receiving echo pulses. It consists of two spark-gap tubes (TR and ATR) that are positioned one-quarter wavelength from the waveguide.

**duty cycle** Duty cycle is the relationship of on-time versus off-time of the pulses. When the duty cycle is increased, the average power output is also increased. Increasing the pulse width and/or pulse repetition rate will increase the duty cycle.

$$\text{Duty cycle} = \frac{\text{Pulse width}}{PRT}$$

where
$PRT$ = Pulse repetition time

**echo box** The echo box is used to tune the radar receiver by providing artificial targets electronically.

**frequency-to-centimeter conversion** While dealing with lower frequency bands, the meter is commonly used to describe wavelength. However, because radar frequencies are much higher, the centimeter (cm) is more easily used. One cm is equal to 0.394 in, and 1 in is equal to 2.54 cm. For example, a radar signal of 3000 MHz would have a wavelength of 10 cm. A convenient method of determining the centimeter wavelength is:

$$\text{Wavelength in centimeters} = \frac{30,000}{\text{Frequency (in MHz)}}$$

**heading flash** The heading flash is a momentary intensification of the sweep line on the PPI presentation whenever the radar antenna points dead ahead. It is actuated by a cam-actuated microswitch in the antenna assembly. When the radar operator knows when the antenna is dead ahead, he or she is able to determine accurately the relative positions of the targets.

**interference to communications receiver** Because of the many harmonics produced by pulsing, radar transmitters can cause interference on most any communication frequency. It will appear as a steady tone corresponding to the pulse rate. Noise or *hash* is produced by a faulty motor generator, or poor grounding or shielding. If interference is present, all units should be checked for proper grounding, bonding, and shielding. Filters can also be installed.

**interference to loran** Radar interference on a loran scope appears as narrow vertical spikes moving across the screen. These are caused by the radar pulses. It is very important to provide proper shielding and grounding between cables, especially long cables. Long connecting lines between the radar transmitter and radar modulator especially need to be grounded and shielded. A second form of interference on the loran scope is called *grass* and is caused by a faulty motor generator unit. Grounding, bonding, commutator, slip rings, and brushes should all be checked.

**intermediate frequency (IF)** The IF frequency in radar units is usually 30 MHz or 60 MHz.

**keep-alive voltage** This voltage is applied to the TR tube to partially ionize it, making it more sensitive to the high power radar transmitter pulses.

**klystron** The kylstron is used as the local oscillator in the radar receiver. Large changes are made in its frequency by changing the volume of its resonant cavity. Smaller frequency changes can be made by controlling the voltage on the repeller plate.

Briefly, klystron is as follows: the cathode emits a stream of electrons. The electrons are directed toward the resonant cavity, and accelerated by the accelerating grid. When the electrons reach the resonant cavity, it is shocked into oscillation at its resonant frequency. This oscillation causes a variation in the potential of the cavity grids (one at the input and one at the output of the cavity). This causes a bunching together of the electrons. As the electron bunches pass through the cavity they encounter the repeller plate, which has a negative potential. The electrons are thus repelled back into the cavity in proper phase, thereby maintaining the oscillations.

**magnetron** The magnetron is the most commonly used radar transmitting tube. It is a type of diode, as it has only a cathode and anode. The magnetron operates as a self-contained oscillator when high voltage pulses are provided to its input. The magnetron construction is begun with a solid circular piece of brass. A large hole is drilled into its center for placement of the cathode. The remaining portion of the circular metal piece is called the anode block. Several smaller holes are drilled into the anode block in such a way that they surround the central cathode. These holes act as resonant cavities and are connected to the central hole by small slots. The resonant cavities function as inductors, and the slots function as capacitors, forming an LC parallel resonant circuit. It is this combination of capacitance and inductance that determines the resonant frequency of the magnetron oscillator. Operation of the magnetron requires a strong external magnetic field supplied by a permanent magnet. If the magnetron is operated without the magnetic field, it will be quickly destroyed by excessive current flow.

**magnetron (service)** A radar service technician should take special precautions when working with a magnetron to prevent damaging it. The magnetron is a delicate electron tube and should be treated accordingly. The magnet should be protected so that its magnetic field will not be weakened. The magnet should not be subjected to extreme heat, shocks, or blows. All magnetic materials such as tools should be kept away from the magnet. The anode is kept at ground potential for safety reasons.

**minimum range**   The minimum range is determined by the time between transmitter pulses. There must be ample time for the radar signal to reach the target and return before the next pulse is sent out.

**pie sections**   Bright flashing pie sections can appear on the radar PPI scope when there is a cracked or defective AFC crystal.

**plan position indicator (PPI)**   The PPI is the cathode-ray tube radar display unit. It provides a map showing the locations of various objects around the ship. The range and azimuth of each object can be read directly off the PPI presentation. As with any CRT, the intensity should be turned down when not in use, or a portion of the phosphor can be damaged.

**pulse repetition rate (PPR)**   The PPR is the number of pulses per second that are applied to the magnetron, and therefore the number of radar pulses that are sent out to the target. The PPR can be adjusted over a wide range of frequencies. Most commonly they range between 400 and 4000 Hz. The higher rates are used with closed targets, but lower rates must be used with distant targets, because more time is needed for the pulses to return. One pulse must return to the receiver before another pulse is sent out.

**pulse repetition time (PRT)**   The PRT is the inverse of PPR. For example, if the pulse repetition rate is 2000 Hz, the pulse repetition time is the reciprocal of 2000 (1/2000) or 0.0005 s.

**radar mile**   A radar mile is the time required for the radar wave to complete a round trip (from the transmitter to the target and back). Because radio waves propagate one nautical mile in 6.2 $\mu$s, a radar mile is 6.2 times 2 or 12.4 microseconds for the round trip.

**radar unit operation and maintenance**   To install, maintain and service a radar unit the operator shall have a first- or second-class radiotelephone or radiotelegraph license with the ship radar endorsement. Log entries are also made under the direct supervision of this licensed person. Replacement of receiving type tubes and fuses can be done by an unlicensed person. The master of the ship, or anyone (unlicensed) designated by the master can operate a ship radar station. However, the equipment shall employ as its frequency-determining element a nontunable, pulse-type magnetron, or other fixed-tuned device. An unlicensed person cannot service, maintain, or be responsible for the radar unit.

**range markers (range rings)**   These are concentric intensified circles on the PPI screen. The distance of objects is read directly off the screen by noting where they are with respect to the range marker. For example, on the 4-mi range, each ring represents 1 mi from the ship (center). On the 1-mi range, each ring stands for 1/4 mi from the ship. The markers are formed by a ringing oscillator with a variable LC circuit. To change from short range to long range, the LC circuit in the oscillator is changed to produce a lower oscillating frequency. A sine wave with a frequency of 80.7 kHz produces a complete cycle in 12.4 $\mu$s, one radar mile (1/0.0807 MHz = 12.4). Therefore, if the LC circuit causes the oscillator to produce 80.7 kHz oscillations, the range rings will represent one radar mile per ring. Range markers are determined by the timer (synchronizer).

**repeller plate (reflector plate)**  The repeller is the negative plate in the klystron tube that reflects the electron beam back into the cavity. By adjusting the voltage potential on this plate, the frequency of the klystron is changed. The AFC error voltage is applied to this repeller plate to control the frequency of the klystron local oscillator.

**safety precautions**  While making repairs or adjustments on radar sets, be sure to discharge all high voltage capacitors with a grounding cable. Use gloves and goggles when working with the CRT. If a CRT is dropped, the implosion can hurt people and damage equipment. Before testing a radar transmitter, it would be a good idea to make sure no explosive or flammable cargo is being handled. RF arcing between metallic surfaces can cause an explosion.

**sea return**  Sea return is the reflection of the transmitted pulses off nearby waves. It is eliminated by the STC circuitry by desensitizing the receiver for nearby objects.

**sensitivity time control (STC)**  The STC circuit automatically reduces the gain of the radar receiver for nearby targets. This desensitizing of the receiver prevents *sea return*, the interference from the reflections of nearby waves. The circuit usually incorporates a hydrogen thyratron tube.

**silicon crystals**  Silicon crystals are used in the mixer and detector stages of the radar receiver. The crystals are very sensitive to static charges. Before handling one, discharge any static to ground. Wrap crystals in lead foil for storage. A high-impedance voltmeter can be used to determine the condition of a crystal. The resistance is measured across the crystal leads, then the voltmeter leads are reversed to measure the resistance in the opposite direction. The larger resistance divided by the smaller resistance gives you the front-to-back ratio. This ratio is normally about twenty to one. A defective crystal in the AFC system will cause bright flashing pie sections to appear on the radar PPI scope. The typical current for a silicon crystal in a radar mixer or detector circuit is 3 mA.

**thyratron tube**  The thyratron is a gaseous vacuum tube. Hydrogen gas is commonly used because of its ability to ionize and deionize rapidly. The hydrogen thyratron is used in the STC circuitry.

**timer**    The timer (synchronizer) circuit has many important functions. It determines the pulse repetition rate and range markers. It provides range markers, sweep, blanking, and unblanking signals for the CRT.

**TR box**    The TR box, or transmit-receive cavity, protects the receiver from the powerful transmitter pulses. It acts as a transmit-receive switch. The TR box contains a gas-filled spark gap tube that ionizes when the transmitter is on. When ionization occurs, the cavity detunes and does not conduct the transmitted pulses to the receiver crystal mixer. When the transmission of the pulse is completed, the spark gap tube deionizes, leaving the receiver ready to receive echo pulses from the target. A dc keep-alive voltage is applied across the TR tube to make it more sensitive. This voltage is slightly below the ionization potential. If the TR box becomes defective, the crystal mixer will be destroyed. This necessitates the replacement of both the TR box and the crystal.

**traveling wave tube (TWT)**    The TWT is a microwave amplifier tube. An electron gun sends a beam of electrons through a helical coil to the anode at the opposite end of the tube. The RF input is inserted at the cathode end of the helix through a coil that surrounds that portion of the tube. The interaction between the electron beam and magnetic field of the helix causes an ever-increasing, bunching-together of electrons as they travel through the tube. By the time they reach the end of the tube, the bunching is maximum. These bunched electrons interact with an output coil, inducing a current. A greatly amplified signal is produced.

**unblanking pulses**    These are pulses produced by the timer and applied to the CRT.

**waveguides**    A waveguide is a transmission line used at microwave frequencies. It consists of a hollow circular or rectangular metal pipe. The rectangular types are preferred because the desired polarization can be maintained. Waveguides are used instead of coaxial cables because their losses are considerably less. Long horizontal runs of waveguide are not desirable because of the likelihood of moisture accumulation in the line. Small holes are sometimes drilled in waveguide in such a way that the condensed moisture is drained.

# Important calculations

## Propagation time of radar pulse

*Problem*    How much time would be required for a radar pulse to travel to an object 10 nautical miles away and return to the radar receiver?

*Solution*    Because one radar mile is 12.4 microseconds, the time required would be 12.4 microseconds times 10 nautical miles, or 124 microseconds. Remember that a radar mile is a round trip mile. In this case it is a total distance of 20 miles.

*Problem*    If the object is 37.2 statute miles away, how much time is required for the radar pulse to travel to the object and return to the radar receiver?

*Solution* The radar wave travels at a rate of 0.186 statute miles in one microsecond. The distance to the object must be doubled because the radar pulse must return to the receiver. Calculate it as follows:

$$\text{time required} = \frac{2 \times \text{distance}}{0.186 \text{ mile}} = 400 \text{ } \mu s$$

*Problem* What is the distance in nautical miles to a target if it takes 310 microseconds for a radar pulse to travel from the radar antenna to the target, and back to the antenna?

*Solution 1* A radar mile is 12.4 microseconds. Because the delay time is 310 microseconds, calculate as follows:

$$\text{distance} = \frac{\text{delay time}}{\text{radar mile}} = \frac{310 \text{ } \mu s}{12.4} = 25 \text{ nautical miles}$$

*Solution 2* Remember that radar ranging is a two-way process. That is, the radar signal must travel to the target and back. It was previously stated that the radar signal travels at the rate of 0.162 nautical miles per microsecond. Therefore, the actual radar range velocity is one half of 0.162 miles per microsecond, or 0.081 nautical miles per microsecond. For example, the radar unit indicates that the round-trip delay time is 310 microseconds, the actual range of the target is $310 \times 0.081$, or about 25 nautical miles.

*Problem* The maximum range of the many ship radar sets is about 40 mi. How long would it take for the radar signal to travel to the object and return?

*Solution* $40 \times 12.4 = 496 \text{ } \mu s$

## Pulse repetition time

Pulse repetition time (PRT) is the reciprocal of the pulse repetition rate (PPR). PRT equals 1/PRR and PRR equals 1/PRT.

*Problem* If the PRR is 2000 Hz, what is the PRT?

*Solution*

$$\text{PRT} = \frac{1}{2000} = 0.005 \text{ s} = 500 \text{ } \mu s$$

## Duty cycle

*Problem* If a radar set has a PPR of 2000 Hz and a pulse width of 0.05 $\mu s$, what is the duty cycle?

*Solution* The PRR must first be converted to PRT as shown above.

$$\text{Duty cycle} = \frac{\text{Pulse width}}{\text{Pulse repetition time}}$$

$$= \frac{0.05 \times 10^{-6}}{0.0005}$$

$$= 0.0001$$

*Problem*  If a radar set has a PPR of 900 and a pulse width of μs, what is the duty cycle?

*Solution*  The PPT is equal to 1/900, or $1.11 \times 10^3$ or 0.0011.

$$\text{Duty cycle} = \frac{\text{Pulse width}}{\text{Pulse repetition time}}$$

$$= \frac{1 \times 10^{-6}}{0.0011}$$

$$= 0.0009$$

## Average power

*Problem*  If a radar set has a PPR of 1000 Hz, a pulse width of μs, and a peak of 100 kW, what is the average power output?

*Solution*  Average power equals peak power times duty cycle. Therefore, the PRT must first be calculated as above. That value is then used to calculate the duty cycle. When the duty cycle is known, simply multiply it by the peak power for the solution.

$$\text{Average power} = \text{Peak power} \times \text{duty cycle}$$

$$= 100{,}000 \text{ W} \times 0.001$$

$$= 100 \text{ W}$$

## Peak power

Peak power is often 5 to 10 kW, while average power is usually only a few hundred watts. This is because the transmitting time (pulse duration) is one or two microseconds. There can be as much a 500 μs of listening time before the next pulse is sent. This is to allow the reflected signal to reach the receiver. Peak power is calculated by transposing the above equation as follows:

$$\text{Peak power} = \frac{\text{Average power}}{\text{Duty cycle}}$$

## Frequency to centimeter conversion

The following formula is a convenient method of determining the wavelength when the frequency in megahertz is known.

$$\text{Wavelength (in centimeters)} = \frac{30{,}000}{\text{Frequency (in megahertz)}}$$

*Problem*  What is the wavelength of a radar signal of 3000 MHz?
*Solution*

$$\text{Wavelength} = \frac{30{,}000}{3000}$$

$$= 10 \text{ cm}$$

*Problem*  How long is a one-half wavelength of the above wave? How long is a one-quarter wavelength of the above?

*Solution* One-half wave would be 10/2 or 5 cm. One-quarter wavelength would be 10/4 or 2.5 cm.

Figure 20-2 illustrates wavelength relationships in a radar system. Assuming the operating frequency is 3000 MHz, the distance to the spark gap is 2.5 cm, or one-quarter wavelength from the center of the waveguide. The distance from the center of the waveguide to the receiver is 5 cm, or one-half wavelength at the operating frequency. These distances are very critical.

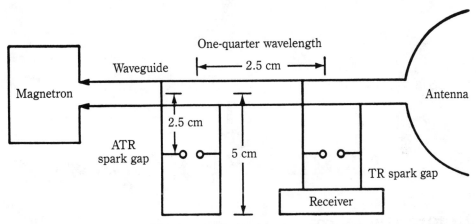

**20-2** Wavelength relationships in radar system.

## Minimum range

The minimum range of a radar unit depends upon the amount of time between transmit pulses. There must be enough *listening time* between pulses to allow the echo to return and enter the receiver. Remember that a radar mile is about 12.4 μs. During that period of time, the pulse travels to the target and back to the receiver, a total of two miles. It is also important to remember that radar signals travel about 164 yards per microsecond one-way distance-to-target, or 328 yards round trip.

# 21
# Exam-taking tips

## Before the exam

- Understand the material in this study guide. Refer to additional texts as necessary.
- Review the material in this study guide. The study questions that follow each chapter will give you a good indication of your knowledge and preparation for the FCC test.
- Get a good night's sleep. Last-minute cramming can be counterproductive.
- Remember to bring your preregistration. It gives the location of the exam as well as the time. The FCC often assigns two or three different exam times during the test date.
- Remember your calculator. The FCC allows most calculators but do not allow hand-held computers. You might want to call the FCC field office and ask them about your particular calculator if you want to be sure.

## At the exam

- Make any necessary bathroom stops before you enter the test area, because the test must be completed before you leave the room.
- After you receive your exam, briefly look it over as you count the pages to make sure they are all present.
- Answer only the questions you feel sure about first. This will encourage you. After you have answered all the easy questions, go back and tackle the more difficult ones. Sometimes an initially difficult question will be clear when you return to it.

- Finally, if you absolutely do not know the answer to a question, use the process of elimination. The following is a suggested method:
  1. List A, B, C, D, and E on your scrap paper.
  2. As you read over the multiple answers, cross off the unlikely answers from your list.
  3. If you can narrow it down to two possible answers, you will have a 50% chance of getting it right when you guess.
- Leave no unanswered questions!

Good Luck!

# 22
# Sample FCC-type practice exam

Please take note, the following practice exams are intended to assist you in finding your weak areas. They are not exact questions from the FCC tests, although some questions are very similar. The FCC exams are considerably more difficult. The FCC uses five possible answers, as opposed to the four in our exams. The FCC wording of test questions is very complex. This means that you must really know the information. Please do not get a false sense of security if you do well on these exams. Have a healthy respect for the FCC test and continue to review the materials in the book.

## Element 3, exam 1

1. Angular modulation includes which of the following?
    a. Amplitude modulation
    b. Frequency modulation
    c. Phase modulation
    d. All of the above
    e. Only B and C

2. The term *azimuth* refers to which of the following?
    a. Vertical distance from the horizon to the zenith
    b. Distance vertically from one horizon through the zenith to the other horizon
    c. Distance from the zenith to the northern horizon
    d. Distance from the zenith to the southern horizon
    e. Horizontal measurement around the 360° horizon

3. What is another way of describing one hertz?

   a. 2 Π diameter
   b. 2 Π radians
   c. 2 Π radius
   d. cycle/minute
   e. volt/meter

4. The binary number 1101 is equal to what decimal number?

   a. 10
   b. 11
   c. 3
   d. 13
   e. 8

5. The glidescope portion of the ILS (instrument landing system) refers to which of the following?

   a. Provides horizontal information relating to aircraft position
   b. Provides vertical positioning information
   c. Provides azimuth information relating to approach
   d. Provides vertical gyro information
   e. Provides horizontal gyro information

6. Which of the following describes the localizer portion of the ILS (instrument landing system)?

   a. Provides vertical position information
   b. Provides horizontal position information
   c. Provides vertical gyro information
   d. Provides horizontal gyro information
   e. None of the above

7. In the IC (integrated circuit), shown in Fig. 22-1, which is pin number 10?

   a. a
   b. b
   c. c
   d. d
   e. e

8. On what frequency does loran C operate?

   a. 2182 kHz
   b. 500 kHz
   c. 100 kHz
   d. 100 MHz
   e. 1000 MHz

Top view

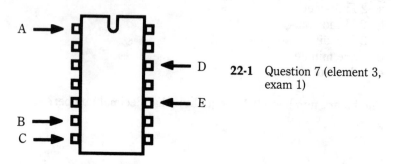

**22-1** Question 7 (element 3, exam 1)

9. One statute mile is equal to how many nautical miles?
   a. 0.868
   b. 0.955
   c. 1.520
   d. 2.520
   e. 1.868

10. If the UTC time is 0200, in the month of December, what time is it in Oregon?
    a. 4 PM
    b. 5 PM
    c. 6 PM
    d. 7 PM
    e. 8 PM

11. What type of EPIRB is required for vessels that travel more than 20 mi off shore?
    a. Class A EPIRB
    b. Class A or class B EPIRB
    c. Class C EPIRB
    d. Class S EPIRB
    e. Any EPIRB

12. Very low frequency (VLF) refers to what frequency band?
    a. 0.3 to 3 kHz
    b. 3 to 30 kHz
    c. 30 to 300 kHz
    d. 3 to 30 MHz
    e. 30 to 300 MHz

13. On what frequencies does the class A EPIRB operate?

    a. 121.5 MHz
    b. 243 MHz
    c. 156.8 MHz
    d. 243 MHz and 121.5 MHz
    e. 121.5 MHz and 156.8 MHz

14. What must you do to ensure proper wetting or solderability of an electronic connection?

    a. Thoroughly clean both surfaces.
    b. Make the connections as tight as possible.
    c. Heat the connections before applying the solder.
    d. Make sure you apply enough heat to ensure free-flowing solder.
    e. All of the above

15. Which type of digital circuit gate does Fig. 22-2 fit?

    a. AND gate
    b. OR gate
    c. NAND gate
    d. NOR gate
    e. NOT gate

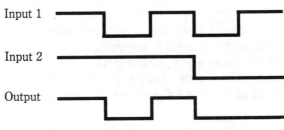

**22-2**   Question 15 (element 3, exam 1)

16. A transmitter requires 200 W of power; a receiver requires 60 W; miscellaneous equipment requires 100 W of power. The supply battery is a 12-V battery with a 60 Ah rating. How long will it take to discharge the battery to 50% of its charge?

    a. 0.5 h
    b. 1 h
    c. 2 h
    d. 3 h
    e. Not enough information given

17. A nonstandard receiver has an incoming frequency of 100 MHz and an image frequency of 140 MHz. What is the IF frequency of the receiver?
    a. 10.7 MHz
    b. 10 MHz
    c. 20 MHz
    d. 40 MHz
    e. Not enough information to calculate

18. In order to obtain maximum antenna radiation to all surrounding points use a:
    a. Parasitic array
    b. Horizontal Hertz
    c. Horizontal Marconi
    d. Vertical Hertz
    e. Vertical Marconi

19. A choke joint is sometimes used over a flanged joint in a waveguide because:
    a. They are more economical.
    b. They offer minimal signal loss at the junction.
    c. They offer maximal attenuation.
    d. They offer poor electrical continuity.
    e. They reflect an electrical short circuit away from the joint.

20. Knife-edge refraction:
    a. Extends the distance of microwave communications.
    b. Is a reduction of atmospheric attenuation of a radio wave when it passes over a sharp object like a mountain ridge.
    c. Attenuates UHF signals during daylight hours.
    d. Is the bending of a UHF signal around a building or sharp edge.
    e. Both b and d.

21. Waveguides should not use which of the following?
    a. Copper materials
    b. Silver lining
    c. Long horizontal runs
    d. Short horizontal runs
    e. Rectangular slots

22. How often must an aviation transmitter located on an airport have its output frequency-checked?
    a. Once per year?
    b. Whenever notified by the FCC of off-frequency operation
    c. Whenever the transmitter is adjusted in the field
    d. If you suspect the transmitter is operating off frequency
    e. All of the above

23. A ship operating a radiotelephone station must monitor what frequency?
    a. 500 kHz
    b. 2182 MHz
    c. 156.8 MHz
    d. 2182 kHz
    e. None of the above

24. A directional wattmeter measures 75 W forward power and 15 W reflected power. What is the "true forward power" or transmitter output power?
    a. 75 W
    b. 90 W
    c. 15 W
    d. 60 W
    e. None of the above

25. The speed of a series wound dc motor varies with:
    a. The number of commutator bars
    b. The number of slip rings
    c. The load applied to the motor
    d. The type of current used
    e. The strength of the interpoles

26. What does the term *bandwidth of emission* include?
    a. All of the output frequencies from a transmitter, regardless of their power levels
    b. Sidebands only
    c. Sidebands and carrier frequencies
    d. Any frequencies that are 0.25% of the total power
    e. Any frequencies that are 0.50% of the total power

27. Why are waveguides not used at frequencies below UHF?
    a. The FCC prohibits it.
    b. The waveguide would be too small and very fragile.
    c. They would require long horizontal runs.
    d. They would be too large to work with.
    e. Low frequency waveguides are commonly used by the highway department as tunnels.

28. How often should the carrier frequency of a crystal-controlled transmitter with an authorized power of 4 W be checked?
    a. Once per week
    b. Once per month
    c. Once per year
    d. Once every 3 years
    e. Whenever it is convenient

29. A transmitter is operating on 2182 kHz. What is its highest harmonic radiation listed below?

    a. 2182 kHz
    b. 4634 kHz
    c. 8782 kHz
    d. 6546 kHz
    e. 10,900 kHz

30. When replacing components of an RF amplifier circuit:

    a. Keep the leads as long as possible
    b. Roll the longer leads into loops so they will take up less room
    c. Keep the leads as short as possible
    d. Use large wire, because of the skin effect
    e. Use only insulated wire

31. How many days in advance must the master or owner of a vessel apply for the annual ship inspection?

    a. 3 days
    b. 7 days
    c. 10 days
    d. 14 days
    e. 21 days

32. What is the purpose of bridge-to-bridge communications (channel 13)?

    a. Only for distress calls
    b. For the listening watch
    c. For weather communications
    d. For navigational communications
    e. For a general calling frequency

33. When is the radiotelephone silent period observed?

    a. Continuously on 2182 kHz
    b. Only when emergency traffic is being conducted
    c. Every 15 min
    d. At the top of each hour and at 30 min after each hour
    e. 15 mins and 45 min after each hour

34. What are the top three priorities of communications?

    a. Distress, safety, then urgency traffic
    b. Urgency, distress, then safety traffic
    c. Distress, urgency, then safety traffic
    d. Distress, urgency, then navigational traffic
    e. Distress, urgency, then rescue traffic

35. The main purpose of the auto alarm system is to attract attention. What three things do auto alarms announce?

    a. That a distress call is about to follow

    b. That a transmission of an urgent cyclone warning is about to follow

    c. The loss of a person overboard

    d. All of the above

    e. None of the above

36. Under what conditions may you exceed the 1-W transmitter limitation in a bridge-to-bridge transmitter?

    a. When rounding the bend in a river

    b. When navigating through a blind spot

    c. When a ship fails to respond to your call on low power

    d. In an emergency situation

    e. In any of the above situations

37. Associated ship units are limited to 1 W of power and may not be transmitted from land. With whom may you communicate, when operating an associated ship unit?

    a. With any other associated ship units within transmitter range

    b. With any ships in the area

    c. With the ship station with which it is associated and with other associated ship units of the same ship

    d. Only with associated units on the same frequency

    e. With no more than three other associated ship units at a time

38. What is the operating frequency of the VHF Omni-range transmitter?

    a. 100 MHz – 200 MHz

    b. 100 kHz – 200 kHz

    c. 2182 kHz

    d. 108 MHz – 118 MHz

    e. 242 MHz – 282 MHz

39. What are the first three letters of the phonetic alphabet?

    a. Alpha, Baker, Charlie

    b. Adam, Bravo, Charlie

    c. Alpha, Bravo, Charlie

    d. Adam, Baker, Charlie

    e. America, Boston, Canada

40. What is the mode of operation that enables two people to talk at the same time?

    a. Simplex

    b. Duplex

  c. Multiplex
  d. Twoplex
  e. Biplex

41. What is the frequency of the auto alarm?

  a. 2182 kHz for both telephone and telegraph
  b. 500 kHz for both telephone and telegraph
  c. 500 kHz for phone and 2182 kHz for telegraph
  d. 2182 kHz for phone and 500 kHz for telegraph
  e. None of the above combinations

42. A transmitter has an output power reading of 100 W, and is connected to 50 feet of antenna transmission line with a power loss of 6 decibels per 100 feet. What is the ERP of the system, if the antenna gain is 3 dB?

  a. 25 W
  b. 50 W
  c. 75 W
  d. 100 W
  e. Not enough information to calculate this

43. A wire has a certain resistance. What would happen to the resistance if it is replaced with a wire of the same length with twice the diameter?

  a. Resistance would be doubled.
  b. Resistance would be halved.
  c. Resistance would be quartered.
  d. Resistance would be quadrupled.
  e. None of the above.

44. What is the frequency of the sine wave in Fig. 22-3, when the time is set at 50 microseconds per division?

  a. 20 kHz
  b. 10 kHz
  c. 4 kHz
  d. 2 kHz
  e. 1 kHz

45. A certain transistor operates as a class B amplifier. At what level is it biased?

  a. 15 V
  b. 7.5 V
  c. 0.7 V
  d. 7 V
  e. 4.6 V

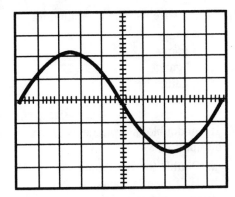

**22-3** Question 44 (element 3, exam 1)

46. The radiotelephone transmitter must be capable of transmitting on 2182 kHz from ship to ship during daytime under normal conditions a minimum range of:

    a. 100 nautical miles
    b. 150 statute miles
    c. 300 nautical miles
    d. 150 nautical miles
    e. 300 statute miles

47. Which of the following is an example of angle modulation?

    a. Amplitude modulation
    b. Frequency modulation
    c. AM vestigial sideband
    d. Single sideband suppressed carrier
    e. Double sideband with full carrier

48. Insulators have what characteristics?

    a. Loosely bound valence electrons.
    b. Tightly bound electrons in the second orbital shell.
    c. Tightly bound protons in the valence shell.
    d. Very few electrons in the valence shell.
    e. Tightly bound electrons in the valence shell.

49. A 12-in. length of wire, having a resistance of 200 $\Omega$ is replaced with a same-length wire having four times the cross-sectional area. What is the resistance of the new piece of wire?

    a. 400 $\Omega$
    b. 200 $\Omega$
    c. 100 $\Omega$
    d. 50 $\Omega$
    e. 25 $\Omega$

50. Which wire has the least amount of electrical resistance?

    a. 6 gauge
    b. 8 gauge
    c. 12 gauge
    d. 4 gauge
    e. 10 gauge

51. If you double the voltage, what must you do to maintain the same amount of power dissipation?

    a. Double the resistance.
    b. Half the resistance.
    c. Quarter the resistance.
    d. Quadruple the resistance.
    e. Do nothing. It will remain the same.

52. If a resistor is banded as red, orange, and yellow, what is its resistance?

    a. 23 Ω
    b. 24 kΩ
    c. 23,000 Ω
    d. 230,000 Ω
    e. 240,000 Ω

53. The sensitivity of a voltmeter is expressed in which of the following terms?

    a. Volts per ohm
    b. Ohms per volt
    c. Ohms per ampere
    d. Volts per volt
    e. Volts per ampere

54. Which of the following is the same as *dc equivalent*?

    a. Average voltage
    b. Peak voltage
    c. Effective current
    d. Peak-to-peak voltage
    e. RMS voltage

55. What are the components of a square wave?

    a. The fundamental frequency plus numerous even harmonics
    b. The fundamental frequency plus numerous odd harmonics
    c. The fundamental frequency plus all the harmonics
    d. Only the even harmonics
    e. Only the odd harmonics

56. Figure 22-4 has a steadily increasing input frequency of a constant amplitude. What happens to the output amplitude?

    a. It remains the same.
    b. It steadily decreases.
    c. It steadily increases.
    d. It increases after resonance is reached.
    e. It decreases after resonance is reached.

**22-4**  Question 56 (element 3, exam 1)

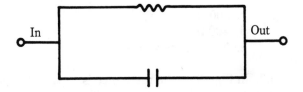

57. In Figure 22-5, which of the following describes the phase relationships of the voltage and the current?

    a. Voltage and current are in phase.
    b. Voltage leads current by 90°.
    c. Voltage leads current by 45°.
    d. Voltage lags current by 90°.
    e. Current leads voltage by 45°.

**22-5**  Question 57 (element 3, exam 1)

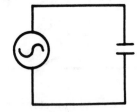

58. Time constant is the time required to charge:

    a. A capacitor up to 36.8% of the supply voltage
    b. A capacitor up to 36.8% of the supply current
    c. A capacitor up to 63.2% of the supply current
    d. A capacitor up to 63.2% of the supply voltage
    e. An inductor up to 63.2% of the supply voltage

59. In Fig. 22-6, the resistance is 100 Ω, the inductive reactance is 200 Ω, and the capacitive reactance is 200 Ω. What is the impedance of this series resonant circuit at resonance?

    a. 100 Ω
    b. 200 Ω

**22-6**  Question 59 (element 3,
exam 1)

c. 300 Ω
d. 400 Ω
e. None of the above

60. In Fig. 22-7, the resistance is 50 Ω, the capacitance is 200 μF and the battery voltage is 1.5 V. How long will it take to fully charge the capacitor?

    a. 10 ms
    b. 20 ms
    c. 30 ms
    d. 40 ms
    e. 50 ms

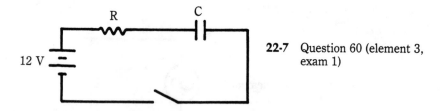

**22-7**  Question 60 (element 3,
exam 1)

61. If an air core of an inductor is replaced with an iron core, how are the characteristics of the inductor changed?

    a. The inductance would decrease.
    b. The inductance would remain unchanged.
    c. The inductance would increase.
    d. The resonant frequency would decrease.
    e. No change would take place.

62. In Fig. 22-8, there are 600 turns on the primary; there are 1800 turns on the secondary; primary voltage is 110 V. What is the current in the secondary if the load resistor is a 3-Ω resistor?

    a. 11 A
    b. 110 mA
    c. 0.011 A
    d. 1100 mA
    e. None of the above

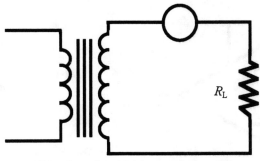

**22-8**   Question 62 (element 3, exam 1)

63. Which of the following is caused by heat dissipation in the iron core of a transformer? (It is sometimes called *iron-core* loss and can be reduced by using a laminated core.)

    a. Hysteresis loss
    b. Copper loss
    c. Eddy current loss
    d. Lamination loss
    e. Core loss

64. In Fig. 22-9, what is the voltage drop across R2, where the battery is 12 V and the resistors are each 100 $\Omega$?

    a. 12 V
    b. 9 V
    c. 6 V
    d. 3 V
    e. 1.5 V

**22-9**   Question 64 (element 3, exam 1)

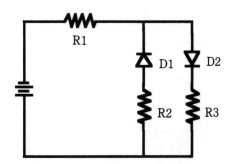

65. In Fig. 22-10, $V_{in}$ is +2 V and $V_{ref}$ is −1 V. And $V_{cc}$ is 6 V. What is the value of the output voltage?

    a. 2 V
    b. −1 V
    c. 4 V

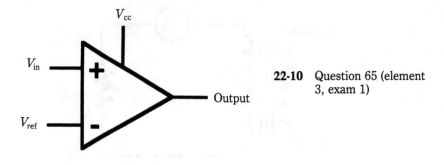

**22-10**   Question 65 (element 3, exam 1)

   d.  5 V
   e.  6 V

66.  In Fig. 22-11, the zener diode is rated at 10 V. What is the output voltage at the load resistor?

   a.  10 V
   b.  9 V
   c.  10.707 V
   d.  10.1414 V
   e.  9.3 V

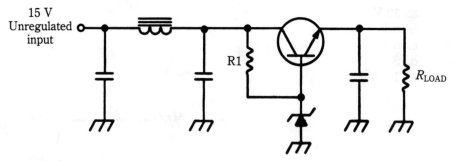

**22-11**   Question 66 (element 3, exam 1)

67.  In Fig. 22-12, $R_1$ is 70 k$\Omega$; $R_2$ is 30 $\Omega$; $E_1$ is 40 V; and $E_2$ is $-30$ V. What is the voltage from point P to ground?

   a.  $-21$ V
   b.  $-12$ V
   c.  12 V
   d.  $-9$ V
   e.  10 V

**22-12**  Question 67 (element 3, exam 1)

68. Which statement is true about Fig. 22-13?

a. Diode D1 is forward biased.
b. Diode D2 is acting like an open switch.
c. R1 and R2 have current flowing through them.
d. R2 and R3 are in parallel, in this circuit.
e. R1 and R3 have current flowing through them.

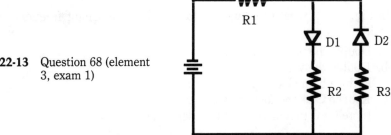

**22-13**  Question 68 (element 3, exam 1)

69. What is the function of the second detector?

a. It demodulates, or detects, and removes the audio signal from the incoming RF signal in FM receivers.
b. It removes all noise components from the signal.
c. It removes the audio from the IF signal in AM receivers.
d. It acts as a mixer for the RF signals.
e. It removes the audio signal from the IF signal in FM receivers.

70. Which type of coupling between receiver stages results in the greatest degree of selectivity (which is the receiver ability to select desired signals and attenuate adjacent signals)?

a. Tight coupling
b. Link coupling
c. Loose coupling

   d. Overcoupling
   e. Narrow-band coupling

71. A pilot is using a transmitter while flying over your home. You can hear him talking on your FM radio on 100.1 MHz. What frequency is he transmitting on?

   a. 100.1 MHz
   b. 110.8 MHz
   c. 121.5 MHz
   d. 132.2 MHz
   e. 50.05 MHz

72. What characteristic of the tank circuit makes it possible to remove the distortions from class B or class C radio frequency amplifiers?

   a. Knife-edge refraction
   b. The distortion effect
   c. The flywheel effect
   d. Rectification
   e. Filtering

73. In Fig. 22-14, what must you do to double the resonant frequency of the tank circuit?

   a. Double the value of L.
   b. Double the value of C.
   c. Reduce both C and L by four times.
   d. Reduce C to one fourth its present value.
   e. Increase C to four times its present value.

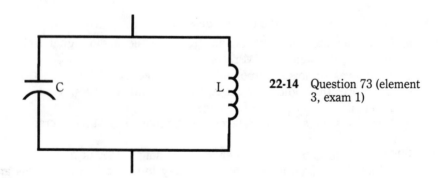

**22-14**  Question 73 (element 3, exam 1)

74. How much power is contained in the upper sideband at 100% modulation, if the power of the unmodulated carrier is 1000 W?

   a. 250 W
   b. 500 W
   c. 750 W

d. 1000 W

e. 1250 W

75. A 200 kHz carrier is modulated with a 4 kHz tone with a strong second harmonic. If the transmitter emission is A3E, what is the bandwidth of emission?

    a. 4 kHz

    b. 8 kHz

    c. 12 kHz

    d. 16 kHz

    e. None of the above

76. Which of the following emissions occupies the most space on the radio spectrum?

    a. 100HA1A

    b. 6K00A3E

    c. 304HF1B

    d. 180KF3E

    e. 3K00H3E

77. In SSSB (also known as J3E), how much carrier suppression, below peak envelope power, is required?

    a. 3 dB

    b. 6 dB

    c. 30 dB

    d. 40 dB

    e. 50 dB

78. Which of the following will increase the resonant frequency of an antenna?

    a. Increase the length of the elements.

    b. Add a series inductor.

    c. Add a series capacitor.

    d. Increase the element thickness.

    e. Add more elements.

79. What is the wavelength of a 500 MHz radio wave?

    a. 0.06 m

    b. 0.6 cm

    c. 600 m

    d. 60 cm

    e. 60 m

80. In Fig. 22-15, what is pictured?

    a. Voltage on a half-wave Hertz antenna

    b. Current on a half-wave Marconi antenna

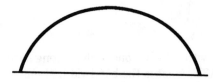

**22-15** Question 80 (element 3, exam 1)

   c. Voltage on a quarter-wave Marconi antenna
   d. Current on a half-wave Hertz antenna
   e. Current on a quarter-wave Hertz antenna

81. How can you increase the reception ability of an antenna?

   a. Increase the number of parasitic elements
   b. By stacking antennas
   c. By increasing the forward gain
   d. By increasing the number of directors
   e. All of the above

82. Which statement is true about coaxial cable?

   a. Some coaxial cable is filled with hydrogen gas to prevent moisture.
   b. Sometimes coax is filled with oxygen gas.
   c. Sometimes coax is filled with nitrogen gas to prevent moisture in the line.
   d. Coax is never filled with gas.
   e. Coaxial cable has two concentric conductors—the center conductor and the shield. The center conductor is usually grounded.

83. When the impedance of a transmission line is properly matched to the antenna, which of the following takes place?

   a. Maximum loss of power
   b. Minimum transfer of power
   c. Maximal transfer of power
   d. A large number of desirable harmonics are produced
   e. Minimum transmission line loss occurs

84. Where is the signal energy coupled in a traveling wave tube?

   a. The plate circuit
   b. The suppressor circuit
   c. The filament circuit
   d. The control grid circuit
   e. The cathode end of the helix

85. An isolator is:

   a. The space in an air-filled coaxial cable
   b. The spaces between waveguide sections
   c. A duplexer

d. A microwave filter
e. A ferrite device that allows microwave energy to pass in one direction with little loss but absorbs power in the reverse direction

86. An RFC is placed between a transmitter and a motor-generator for what useful purpose?

a. To provide RF feedback
b. To encourage RF feedback
c. To prevent insulation breakdown
d. To increase the line frequency
e. None of the above

87. Which of the following is used to step up dc voltage?

a. Transformer
b. Dynamotor
c. Motor generator
d. Rectifier
e. Bridge rectifier supply

88. A durable nameplate shall be mounted on the required radiotelephone transmitting and receiving equipment. What information must the nameplate contain?

a. The name of the manufacturer
b. The type or model number
c. The output power rating of the transmitter
d. All of the above
e. Only a and b

89. What precautions must be taken while soldering to ensure permanent solder junctions?

a. Clean connections
b. Good mechanical bond
c. Sufficient heat for adequate wetting
d. Heatsink when indicated
e. All of the above

90. Which of the following statements is true concerning Fig. 22-16?

a. The capacitors protect the diodes from excessive PIV.
b. The capacitors equalize the voltage drop across the diodes.
c. The capacitors are used for filtering purposes.
d. The capacitors protect the diodes from voltage transients.
e. None of the above.

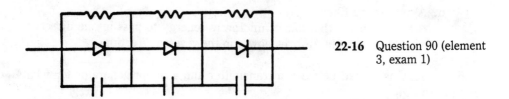

**22-16** Question 90 (element 3, exam 1)

91. When transmitting single-sideband emission, the carrier frequency shall be maintained within what applicable number of Hertz per second of the specified carrier frequency? (for other than Civil Air Patrol stations)

    a. 5 Hz
    b. 10 Hz
    c. 15 Hz
    d. 20 Hz
    e. 30 Hz

92. Convert 0.0025% to parts per million (ppm).

    a. 2.5 ppm
    b. 25 ppm
    c. 250 ppm
    d. 2500 ppm
    e. None of the above

93. Using only a voltmeter, calculate the power dissipated in Fig. 22-17?

    a. Measure voltage W to X, square and divide by 152,000.
    b. Measure voltage Y to Z, square and divide by 2000.
    c. Measure voltage W to X, divide by 150,000, square and multiply by 152,000.
    d. Measure voltage Y to Z, divide by 152,000.
    e. Measure voltage W to Z, divide by 152,000.

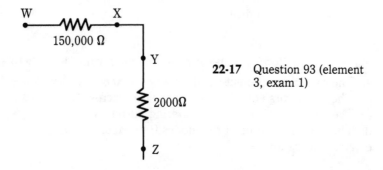

**22-17** Question 93 (element 3, exam 1)

94. What is the purpose of C7 in Fig. 22-18?

  a. Loading
  b. Tuning
  c. Passing dc to the load
  d. Blocking ac to the load
  e. Adjusting internal capacitance

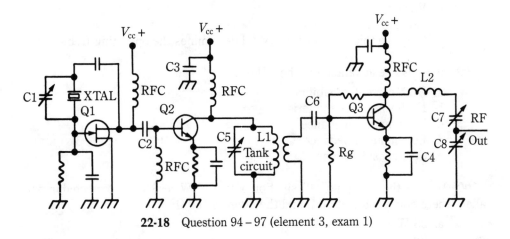

**22-18** Question 94 – 97 (element 3, exam 1)

95. What is the purpose of C8 in Fig. 22-18?

  a. Loading
  b. Tuning
  c. Passing dc to the load
  d. Blocking ac to the load
  e. Adjusting internal capacitance

96. Which of the following components, if it fails (opens), will cause the least amount of damage to the circuit in Fig. 22-18?

  a. Q1
  b. C5
  c. C6
  d. C3
  e. C7

97. What happens to the circuit when the emitter bypass capacitor C4 becomes open? (Fig. 22-18)

  a. Increased gain and output
  b. Reduced gain and output
  c. Circuit will not operate

  d.  No change in circuit
  e.  Q3 would immediately fail

98.  The thickness is the primary factor in determining the resonant frequency of the crystal. What else has a strong influence on the crystal operating frequency?

  a.  Electric fields
  b.  Magnetic fields
  c.  Vibrations
  d.  Temperature changes
  e.  Only the thickness of the crystal determines the operating frequency.

99.  What is one advantage of the JFET?

  a.  High current capability
  b.  High voltage characteristics
  c.  High input impedance
  d.  Low input impedance
  e.  Cost considerations

100.  What is the total power dissipation of two 10-W resistors, connected in parallel, where one is rated at 50 $\Omega$ and the other is 100 $\Omega$?

  a.  25 W
  b.  20
  c.  15
  d.  10
  d.  5

# Element 3, Exam 2

1.  Ship station log books containing communications that are under FCC investigations must be retained for how long?

  a.  1 year
  b.  3 years
  c.  5 years
  d.  7 years
  e.  Until the FCC notifies the ship station licensee in writing that the logs may be destroyed.

2.  How does a wattmeter measure true power?

  a.  By current squared times resistance
  b.  Voltage times current
  c.  By squaring the voltage and dividing by the resistance
  d.  Voltage times current, correcting for phase differences
  e.  The wattmeter does not measure true power.

3. The inductance of a coil varies with:
   a. The cross-sectional area of the wire
   b. The inverse of the number of turns
   c. The square of the number of turns
   d. The square of the core diameter
   e. The number of turns

4. When should the EPIRB battery be replaced?
   a. When 30% of its useful life has expired, as determined by the manufacturer
   b. When 50% of its useful life has expired, as determined by the manufacturer
   c. When 75% of its useful life has expired, as determined by the manufacturer
   d. When 50% of its useful life has expired, as determined by the manufacturer, or after being used in an emergency situation
   e. Whenever the battery does not fully charge, in a reasonable period of time, as determined by the battery manufacturer

5. How does the bridge-to-bridge operator perform station identification?
   a. With the operator's name
   b. With the assigned ship call sign
   c. With the name of master of the vessel
   d. With the name of the vessel, in lieu of the call sign
   e. Any of the above is authorized

6. What is the legal range of transmitter modulation for ship station transmitters?
   a. Between 60% and 100%
   b. Between 75% and 110%
   c. Between 75% and 100%
   d. Between 80% and 110%
   e. Between 80% and 100%

7. What is J3E emission?
   a. Double sideband with carrier suppressed to 40 dB
   b. Single sideband with carrier suppressed to 40% below peak envelope power
   c. Full carrier with a single sideband suppressed to 40 dB below peak envelope power
   d. Single sideband with carrier suppressed to 40 dB below peak envelope power

e. Full carrier with both sidebands suppressed to an appropriate level to meet the technical standards specified in the Code of Federal Regulations

8. Stacking of antennas has what effect?

a. Improves the reception
b. Reduces the reception
c. Decreases the antenna current
d. Decreases directivity
e. Has no effect on reception or directivity

9. To correct split tuning:

a. Use a high $L/C$ ratio
b. Use a low Q circuit
c. Realign the detector stage
d. Increase the coupling between stages
e. Decrease the coupling between stages

10. Residual magnetism can be defined as:

a. Retentivity of the lines of magnetizing force
b. A term applied to intense magnetic fields
c. Polarized molecular alignment in a magnetized material while not under the influence of a magnetizing force
d. A strong permanent magnet
e. None of the above

11. Which section in an AM superheterodyne receiver produces the most noise?

a. Audio amplifier
b. IF amplifier
c. Local oscillator
d. Mixer
e. Detector

12. A class C amplifier is best known for its:

a. Use in audio amplifiers
b. High efficiency
c. Good reproductivity of the input waveform
d. Low degree of distortion
e. Long duty cycle

13. Which of the following contains only components of a 25 kHz square wave?

a. 12.5 kHz, 25 kHz, 75 kHz, 125 kHz
b. 25 kHz, 75 kHz, 100 kHz, 175 kHz
c. 25 kHz, 125 kHz, 175 kHz, 250 kHz

d. 25 kHz, 75 kHz, 125 kHz, 225 kHz

e. None of the above

14. A ship going to sea using 1605 kHz to 4500 kHz radiotelephone equipment must maintain a continuous listening watch on:

a. 1882 kHz

b. 2184 kHz

c. 2182 kHz

d. 2192 kHz

e. 4182 kHz

15. Why is nitrogen gas often pumped into transmission lines and waveguides?

a. To prevent moisture from entering

b. To prevent oxidation of center conductors

c. To act as a conductor

d. To increase the dielectric properties of the transmission line or waveguide, thus reducing the line loss

e. To increase the useful operating frequency, by reducing the line attenuation

16. Channel 16, the distress and calling frequency, operates on what frequency?

a. 158.6 MHz

b. 2182 kHz

c. 168.6 MHz

d. 156.8 kHz

e. 156.8 MHz

17. What channel is used for contacting the U.S. Coast Guard?

a. Channel 12

b. Channel 13

c. Channel 16

d. Channel 20

e. Channel 22

18. If a person purchases an aircraft, having an aircraft station license, what must he or she do about the station license?

a. Transfer the existing license to his or her name

b. Operate on the existing license until he or she gets his own license, providing the purchaser applies immediately

c. Operate without a license

d. Use the same call letters, providing the identification is followed with the name

e. The buyer must apply for his or her own license. The existing license cannot be used because licenses may not be assigned or transferred to another individual.

19. What channel is used for bridge-to-bridge navigational communications?

    a. Channel 6
    b. Channel 12
    c. Channel 13
    d. Channel 15
    e. Channel 22

20. What are the operating frequencies of emergency locator transmitters (ELTs)?

    a. 2182 kHz
    b. 500 kHz
    c. 121 MHz and 253 MHz
    d. 158.6 MHz
    e. They operate on the same frequencies as the class A and class B EPIRBs.

21. When may EPIRB transmitters be checked, when US Coast Guard involvement is not possible?

    a. Only during the first five minutes of each hour
    b. At the top of the hour for three minutes and at thirty minutes after the hour for three minutes
    c. During the first five minutes of any hour for one second or three audio sweeps
    d. They may be tested any time, as long as the technician announces the words *testing, testing, one, two, three.*
    e. They may never be tested without the cooperation of the US Coast Guard.

22. What type of EPIRB is required on survival crafts?

    a. Class A EPIRB
    b. Class B EPIRB
    c. Class C EPIRB
    d. Class S EPIRB
    e. 406 MHz Satellite EPIRB

23. Which of the following are required by the FCC to have the General Radiotelephone Operator License?

    a. Ship station
    b. Ambulance
    c. Police
    d. Television station
    e. All of the above

24. When is the silent period for radiotelephone stations?

    a. For the first five minutes of any hour
    b. For five minutes at the top of the hour and for five minutes thirty minutes after the hour
    c. The silent period is any time the receiver is turned off
    d. The silent period is continuous 2182 kHz
    e. For three minutes starting at the top of each hour and for three minutes starting at thirty minutes after each hour

25. Which of the following is true about Doppler radar?

    a. An approaching target returns a lower frequency echo
    b. A departing target returns a higher frequency echo
    c. A departing target returns a signal equal to the radar transmitter frequency
    d. A departing target returns a signal higher than the radar transmitter frequency
    e. None of the above

26. A radar mile is which of the following:

    a. 12.4 ms—the time required for a radar pulse to travel one nautical mile
    b. 12.4 $\mu$s—the time required for a radar pulse to travel one statute mile
    c. 12.4 $\mu$s—the time required for a radar pulse to travel one nautical mile
    d. 12.4 ms—the time for a radar pulse to travel one nautical mile to a target and back to the radar receiver
    e. 12.4 $\mu$s—the time required for a radar pulse to travel one nautical mile to a target and back to the radar receiver

27. Signal-to-noise ratio can be increased by which of the following?

    a. By using a high percentage of modulation in the transmitter
    b. By using pre-emphasis in the transmitter
    c. By using de-emphasis in the FM receiver
    d. By using a receiver with a narrow bandpass
    e. All of the above

28. What is the upper end of the VLF band?

    a. 0.03 kHz
    b. 0.3 kHz
    c. 3 kHz
    d. 30 kHz
    e. 300 kHz

29. What is the emission designation for amplitude modulation with full carrier and two sidebands?

    a. A3A

    b. A3T

    c. A3E

    d. J3E

    e. H3E

30. Which of the following is a method of telecommunication where a fixed image is converted into a permanent record at the receiving station?

    a. C3F

    b. F3C

    c. R3E

    d. R3P

    e. J3E

31. When a capacitor discharges through a resistor, at what point is the charge on the capacitor equal to 36.8% of its full charge voltage?

    a. After one time constant

    b. After two time constants

    c. After three time constants

    d. After four time constants

    e. After five time constants

32. What is the effective radiated power of a system where the transmitter output power is 100 W; an in-line wattmeter has a loss of 2.5 dB; 50 ft. of transmission line is used that is rated 4 dB loss per 100 ft.; and the antenna gain is 21.2 dB?

    a. 500 W

    b. 5 kW

    c. 50 W

    d. 750 W

    e. Impossible to determine with existing information

33. The field strength of a transmitted signal is 1 mV/m at a certain distance from the transmitter. If $x$ is the transmitter power, how many times would you have to increase it to measure a 4 mV/m field strength at the same distance?

    a. 2 times

    b. 4 times

    c. 8 times

    d. 16 times

    e. 32 times

34. If transmitter power is increased by a factor of four times, how will the field strength at a distance of 1 mi change?

    a. It will increase by 2 times
    b. It will increase by 4 times
    c. It will increase by 6 times
    d. It will increase by 8 times
    e. It will increase by 12 times

35. If the forward power from a transmitter is 100 W and the reflected power from a 6-dB gain antenna is 20 W, what is the effective radiated power of the system?

    a. 480 W
    b. 400 W
    c. 320 W
    d. 150 W
    e. 100 W

36. What type of output does the flip-flop have?

    a. Square wave
    b. Spike wave
    c. Sine wave
    d. Triangular wave
    e. Sawtooth wave

37. Which of the following describes the two outputs of the flip-flop?

    a. Both outputs are sometimes high
    b. Both outputs are sometimes low
    c. The outputs constantly switch between both being high to both being low
    d. One output is high (flip) and the other output is low (flop). By monitoring either output, you will see it change from flip to flop to flip, etc. The two outputs are always opposites.
    e. It does not matter because since the advent of computers, flip-flops are rarely ever used.

38. What is the phonetic alphabet for the letters G, H, I, and J?

    a. George, Hotel, India, John
    b. Golf, Hotel, Indiana, Juliett
    c. Golf, Hotel, India, John
    d. Golf, Henry, Indiana, John
    e. Golf, Hotel, India, Juliett

39. 40 °C is equal to what Fahrenheit temperature?

    a. 8°
    b. 48°

c. 80°
d. 104°
e. 57.6°

40. If the incoming signal to a receiver is 156.8 MHz and the local oscillator frequency is 166.3 MHz, what frequency could potentially cause image reception?

  a. 167.5 MHz
  b. 178.2 MHz
  c. 611.8 MHz
  d. 1066.8 MHz
  e. 175.8 MHz

41. If 50 MHz is applied to the input of a trippler, what is the output of that circuit?

  a. 25 MHz
  b. 50 MHz
  c. 100 MHz
  d. 150 MHz
  e. 200 MHz

42. Atmospheric noise is not a major problem to radiocommunications:

  a. Below 30 kHz
  b. Above 30 MHz
  c. Above 30 kHz
  d. Above 1 MHz
  e. Above 10 MHz

43. The 2182 kHz distress frequency is in what frequency range?

  a. Low frequency (LF)
  b. Medium frequency (MF)
  c. Very high frequency (VHF)
  d. Ultra high frequency (UHF)
  e. Extremely low frequency (ELF)

44. What is the current and voltage relationship in a one-half wave Hertz antenna?

  a. Current is maximum and voltage is minimum at the center
  b. Current is minimum and voltage is maximum at the ends
  c. Current and voltage are maximum at the center
  d. Current and voltage are maximum at the ends
  e. Both A and B are correct

45. Voltage may also be called:

    a. Electromotive force
    b. Difference of potential
    c. *IR* drop
    d. EMF
    e. All of the above

46. The buffer amplifier produces:

    a. A steady light load for the final amplifier
    b. A heavy load for the oscillator
    c. A variable load for the oscillator
    d. Protection for the transmitter
    e. A steady light load for the oscillator

47. A transmitter requires 350 W of power and a receiver requires 50 W of power from a 12-V, 50 Ah battery source. How long would it take to discharge the battery?

    a. 6 h
    b. 3 h
    c. 8 h
    d. 33.3 h
    e. 1.5 h

48. What is the phase angle in a circuit where the capacitive reactance is 70 $\Omega$, the inductive reactance is 30 $\Omega$, and the resistance is 40 $\Omega$?

    a. 90°
    b. 180°
    c. 270°
    d. 45°
    e. 0°

49. What type of current does an ac ammeter indicate?

    a. Peak-to-peak
    b. Amp-to-amp
    c. Average
    d. Peak
    e. Effective

50. Application for annual safety inspections shall be filed by the master of the vessel, owner of the vessel or the ship station licensee how many days in advance?

    a. 3 days in advance
    b. 5 days in advance

    c. 7 days in advance

    d. 14 days in advance

    e. 21 days in advance

51. How long would it take for the capacitor to charge up to the 12-V supply voltage in a circuit where the capacitor is 0.01 $\mu$F and the resistor is 2 M$\Omega$?

    a. 0.02 s

    b. 0.2 s

    c. 0.1 s

    d. 0.01 s

    e. 1 s

52. The auto alarm announces:

    a. A distress call or message is about to follow

    b. An urgent cyclone warning

    c. A loss of a person overboard

    d. Any of the above

    e. A and C only

53. Who sets the frequency standards?

    a. The FCC

    b. The National Bureau of Standards

    c. WWV

    d. The White House

    e. The International Bureau of Standards

54. The radiotelephone distress signal is:

    a. SOS

    b. Distress

    c. Mayday

    d. Help

    e. Emergency

55. Which class of amplifier is the most efficient?

    a. Class A

    b. Class B

    c. Class C

    d. Class D

    e. Class AB

56. A radio operator carrying an associated ship unit may communicate with who?

    a. Any ship that is on the same frequency

    b. Amateur radio operators

c. Other CB operators

d. The ship station with which they are associated or with other associated units of the same ship

e. With associated units of other ships or with bridge-to-bridge stations

57. Ship stations using radiotelephony shall identify by announcing in the English language:

a. The name of the operator

b. The name of the ship

c. The station-assigned call sign

d. The name of the master of the ship

e. The name of the vessel's owning agency

58. Associated ship units must be able to transmit on:

a. Channel 13 and at least one appropriate intership frequency

b. Channel 16 and at least one appropriate intership frequency

c. Channel 22 and at least one appropriate intership frequency

d. Channel 16 only

e. Channel 13 only

59. Of the three basic transistor amplifier configurations, which produces a phase reversal of the input signal?

a. Common base

b. Common collector

c. Common emitter

d. All of the above

e. None of the above

60. After receiving a notice of violation, the operator must make written response within:

a. 3 days

b. 5 days

c. 10 days

d. 14 days

e. 30 days

61. In a TWT (traveling wave tube), the input signal is applied:

a. At the anode end of the helix

b. To the cathode end of the tube

c. At the cathode end of the helix

d. To the anode

e. To the grid of the tube

62. In a varactor:

    a. The capacitance remains constant.
    b. The capacitance varies as the reverse bias of the diode varies.
    c. The resistance varies as the reverse bias across the diode varies.
    d. The voltage varies as the capacitance is changed.
    e. The varactor is a large rheostat that enables a technician to dial in any-where from 0 V to about 140 V.

63. When an A3E transmitter is 100% modulated, how much power is contained in the two sidebands?

    a. $1/6$ of the total power
    b. $1/3$ of the total power
    c. $1/4$ of the total power
    d. $1/2$ of the total power
    e. $1/8$ of the total power

64. Antenna lights shall be checked:

    a. Hourly
    b. Daily
    c. Weekly
    d. Monthly
    e. Yearly, unless you have reason to believe a light might be malfunctioning

65. Station logs containing distress communications shall be maintained for a period of:

    a. 6 months from the date of entry
    b. 1 year from the date of entry
    c. 3 years from the date of entry
    d. 5 years from the date of entry
    e. Until the FCC notifies the operator, in writing, that they may be destroyed

66. If the cross-sectional area of a conductor is tripled, the resistance would become:

    a. Three times
    b. One third
    c. Nine times
    d. One ninth
    e. Six times

67. Push-pull operation:

    a. Eliminates odd harmonics
    b. Eliminates even harmonics
    c. Eliminates all harmonics

    d. Generally has more harmonics than a single-ended amplifier

    e. Amplifies the positive and negative portions of the input cycle plus all harmonics present

68. When a 50-$\Omega$ and a 100-$\Omega$ resistor are connected in parallel, how much power can they safely dissipate when both resistors are rated 10 W?

    a. 5 W

    b. 10 W

    c. 15 W

    d. 20 W

    e. 25 W

69. Selectivity of a receiver is:

    a. The ability to select undesirable signals and amplify them

    b. The ability to select undesirable signals and pass them through appropriate filters to attenuate them

    c. The ability to select wanted signals and attenuate adjacent signals

    d. Greatest when the bandwidth is wide and the circuit Q is low

    e. Greatest when the coupling between stages is tight

70. To double the resonant frequency of an LC tank circuit:

    a. Use twice as much capacitance

    b. Use one half as much capacitance

    c. Use one fourth as much capacitance

    d. Use four times as much capacitance

    e. Use four times as much inductance

71. When impedance is properly matched between transmission line and antenna:

    a. Maximum power will be transferred to the antenna.

    b. Minimum power will be transferred to the antenna.

    c. A high SWR will be present.

    d. A high degree of reflected power will be present.

    e. Both a and c are correct.

72. Which statement is correct about transistor amplifiers?

    a. The emitter-base junction is reverse biased.

    b. The collector-base junction is reverse biased.

    c. The collector is forward biased.

    d. Both junctions are forward biased.

    e. Both junctions are reverse biased.

73. How can you lower the resonant frequency of a Hertz antenna?

    a. Place a capacitor in series with the antenna.

    b. Place a resistor in series with the antenna.

c. Place an inductance in series with the antenna.
d. Place an inductance in parallel with the antenna.
e. Make the antenna physically shorter.

74. Service and maintenance of avionics or marine transmitting equipment:
    a. May be done by anyone who understands electronics
    b. Can be done by anyone appointed by the master of the ship
    c. Must be performed under the immediate supervision and responsibility of a person holding a General Radiotelephone Operator License
    d. Must be conducted only at the factory
    e. Must be conducted by a qualified FCC representative

75. The listening watch on 2182 kHz is to be:
    a. At the top of the hour and at 30 min after the hour for three minutes
    b. At 15 min after the hour and 45 min after the hour for 3 min
    c. Continuous and at all times the station is not being used for authorized traffic
    d. Four times per hour
    e. There is no listening watch on that frequency

76. What is wrong with the circuit in Fig. 22-19?
    a. Emitter-base is reverse biased.
    b. Collector-base is forward biased.
    c. Emitter bias voltage is too high.
    d. Collector bias voltage is too high.
    e. Nothing is wrong with the circuit.

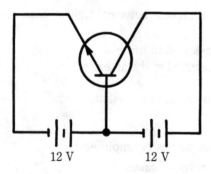

**22-19**  Question 76 (element 3, exam 2)

12 V        12 V

77. What is the total capacitance in Fig. 22-20, when each capacitor has a value of 8 μF.
    a. 4.8 μF
    b. 2.4 μF
    c. 9.6 μF
    d. 32 μF
    e. None of the above

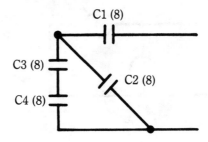

**22-20**  Question 77 (element 3, exam 2)

78. What is the total inductance in Fig. 22-21?
    a. 3.33 H
    b. 6.66 H
    c. 12.33 H
    d. 60 H
    e. None of the above

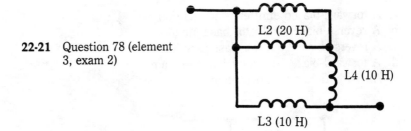

**22-21**  Question 78 (element 3, exam 2)

79. A transmitter has a carrier frequency of 10 MHz and an allowable tolerance of 20 ppm. How much may the crystal oscillator deviate if it is one eighth the frequency of transmitter output?
    a. 2.5 Hz
    b. 25 Hz
    c. 200 Hz
    d. 250 Hz
    e. None of the above

80. Which of the arrangements, in Fig. 22-22 is used to measure power?
    a. Arrangement A
    b. Arrangement B
    c. Arrangement C
    d. Arrangement D
    e. None of the arrangements

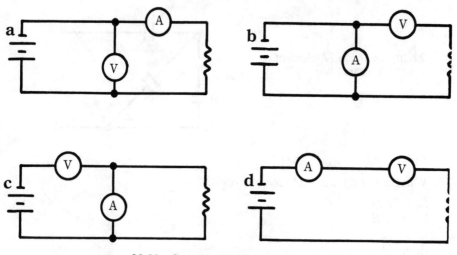

**22-22** Question 80 (element 3, exam 2)

81. What does Fig. 22-23 illustrate?

    a. A forward-biased emitter base junction
    b. A reverse-biased collector base junction
    c. A reverse-biased emitter base junction
    d. A forward-biased collector base junction
    e. An NPN field-effect transistor with proper bias

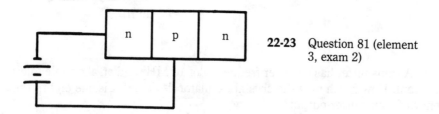

**22-23** Question 81 (element 3, exam 2)

82. Electrical energy is measured by what unit?

    a. Watt
    b. Kilowatt
    c. Kilowatt-hour
    d. Kilovolt
    e. Kilovolt-hour

83. Which of the following is used to step up dc voltage?

    a. Transformer
    b. Dynamotor
    c. Motor-generator

    d. Rectifier supply

    e. Doubler

84. In an electronic circuit at resonance:

    a. Inductive reactance is usually more than capacitive reactance

    b. Capacitive reactance equals and cancels inductive reactance

    c. Resistance equals capacitance

    d. Resistance and inductance and capacitance are all equal

    e. Phase angles are 45°

85. As the gauge of wire becomes smaller, the resistance of the wire:

    a. Increases

    b. Decreases

    c. Remains the same

    d. Increases with the square of the diameter

    e. Increases with the square of the cross-sectional area

86. In a receiver, the limiter:

    a. Removes static and provides a low gain and constant output

    b. Removes the FM from the signal

    c. Removes the AM from the signal

    d. Provides a high gain and variable output

    e. Limits the audio gain in such a way that the volume always remains the same

87. To maintain stability in a crystal oscillator:

    a. Maintain a constant humidity

    b. Maintain a negative bias

    c. Maintain a positive bias

    d. Maintain a constant temperature

    e. Maintain a constant pressure on the crystal

88. A carbon microphone operates on the principle of:

    a. Varying capacitance

    b. Varying resistance

    c. Varying inductance

    d. Varying magnetism

    e. None of the above

89. If the field coil in a shunt-wound motor develops an open, the motor will:

    a. Slow down and destroy itself

    b. Speed up and destroy itself

    c. Stop

    d. Slow down to a slower speed

    e. Not be affected

90. To determine if an RF amplifier has been properly neutralized, use a:
    a. VTVM
    b. Neon bulb
    c. Milliammeter
    d. None of the above
    e. RF amplifiers do not need to be neutralized.

91. An audio amplifier has an output of 6000 Ω. What is the transformer ratio (primary to secondary) if it is to be connected to a 15-Ω speaker?
    a. 400:1
    b. 4:1
    c. 20:1
    d. 1:20
    e. 25:1

92. The purpose of auto alarm signal is to:
    a. Test the transmitter
    b. Alert the crew of fire in the engine room
    c. Alert the crew of a burglary
    d. Attract attention
    e. Alert the crew of shallow water

93. What is the series impedance in a circuit where the resistance is 1 MΩ, the capacitance is 0.001 $\mu$F, and 2 $\pi$F is 1000?
    a. 1 MΩ
    b. 2 MΩ
    c. 1.414 MΩ
    d. 1.212 MΩ
    e. 4 MΩ

94. In Fig. 22-24, with S1 and S2 off, all meters show zero current flow. What happens to the current flow through the meters when S1 is turned on?
    a. All three meters show current flow.
    b. M1 and M3 show current flow.
    c. Only M1 shows current flow.
    d. Only M1 and M2 show current flow.
    e. Nothing happens because both S1 and S2 must be on for the circuit to operate.

95. The statute mile is equal to how many nautical miles?
    a. 0.686 nautical miles
    b. 0.868 nautical miles
    c. 1.151 nautical miles
    d. 1.515 nautical miles
    e. 1 nautical mile

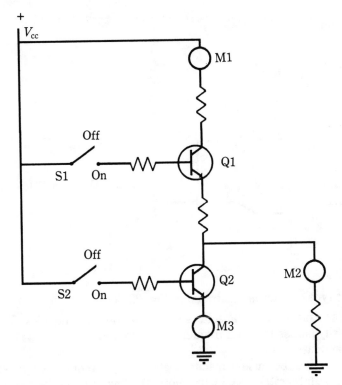

**22-24**   Question 94 (element 3, exam 2)

96. A transmitter has a carrier frequency of 10 MHz and an allowable frequency tolerance of 0.025%. What is the maximal allowable deviation from center frequency?

    a. 2500 kHz
    b. 250 kHz
    c. 25 kHz
    d. 2.5 kHz
    e. 250 Hz

97. How is the gain calculated in the op-amp circuit shown in Fig. 22-25?

    a. $\text{Gain} = \dfrac{R_f}{R_i}$

    b. $\text{Gain} = \dfrac{R_i}{R_f}$

    c. $\text{Gain} = -\dfrac{R_f}{R_i}$

d. Gain $= -\dfrac{R_i}{R_f}$

e. Gain $= \dfrac{R_f}{R_i} + 1$

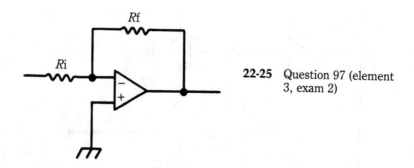

**22-25** Question 97 (element 3, exam 2)

98. What type of filter is illustrated in Fig. 22-26?
    a. Low-pass because capacitors easily pass low frequencies. The inductor easily passes the high frequencies to ground.
    b. Bandpass
    c. High-pass because capacitors allow high frequencies to easily pass through them. The inductor shorts the low frequencies to ground.
    d. Band stop filter
    e. Capacitive input filter

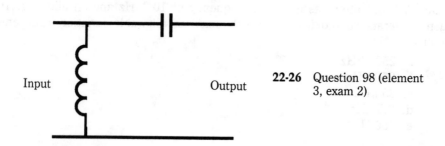

**22-26** Question 98 (element 3, exam 2)

99. What input condition would allow a high output condition in Fig. 22-27?

| | K | L | M | N |
|---|---|---|---|---|
| a. | 1 | 1 | 1 | 1 |
| b. | 1 | 1 | 1 | 0 |
| c. | 1 | 1 | 0 | 1 |
| d. | 1 | 1 | 0 | 0 |
| e. | 0 | 0 | 0 | 1 |

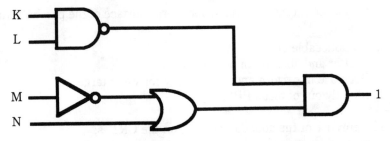

**22-27**   Question 99 (element 3, exam 2)

100.  What is the total capacitance between points A and B in Fig. 22-28? (All capacitors are 2 μF.)

   a.  0.5 μF
   b.  0.8 μF
   c.  2 μF
   d.  5 μF
   e.  8 μF

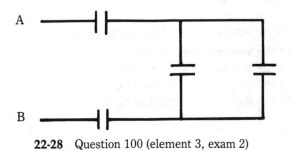

**22-28**   Question 100 (element 3, exam 2)

# Element 8, exam 1

1.  One radar mile is how many microseconds?


   a.  6.2
   b.  12.4
   c.  528.0
   d.  .186


2.  Good bearing resolution largely depends upon:
   a.  A high pulse repetition rate
   b.  A high duty cycle
   c.  A narrow antenna beam in the vertical plane
   d.  A narrow antenna beam in the horizontal plane

3. A thick layer of crust and corrosion on the surface of the parabolic dish will have what effect?

    a. No noticeable effect
    b. Scatter and absorption of radar wave
    c. Decrease in performance, especially for weak targets
    d. Slightly out of focus PPI scope

4. The purpose of the aquadag coating on the CRT is:

    a. To protect the electrons from strong electric fields
    b. To act as a second anode
    c. To attract secondary emission from the CRT screen
    d. To give a final acceleration to the electron beam
    e. All of the above

5. An artificial transmission fine is used for:

    a. Transmission of radar pulses
    b. For testing the radar unit, when actual targets are not available
    c. Determining shape and duration of pulses
    d. Testing the delay time for artificial targets

6. Where is an RF attenuator used in a radar unit?

    a. Between the antenna and the receiver
    b. Between the magnetron and the antenna
    c. Between the magnetron and the AFC section of the receiver
    d. Between the AFC section and the klystron

7. The ATR box:

    a. Protects the receiver from strong radar signals
    b. Prevents the received signal from entering the transmitter
    c. Turns off the receiver when the transmitter is on
    d. Both a and c are correct

8. Which of the following is one of the authorized frequency bands for ship radar?

    a. 2900 – 3200 MHz
    b. 9100 – 9300 MHz
    c. 5460 – 5650 MHz
    d. 4460 – 4650 MHz

9. The AFC system is used to:

    a. Control the frequency of the magnetron
    b. Control the frequency of the klystron
    c. Control the receiver gain
    d. Control the frequency of the incoming pulses

10. Bearing resolution is:
    a. The ability to distinguish two targets of different distances
    b. The ability to distinguish two targets of different elevations
    c. The ability to distinguish two adjacent targets of equal distance
    d. The ability to distinguish two targets of different size

11. In the AFC system, the discriminator compares the frequencies of the:
    a. Magnetron and klystron
    b. PRR generator and magnetron
    c. Magnetron and crystal detector
    d. Magnetron and video amplifier

12. What frequency is the discriminator tuned to?
    a. The magnetron frequency
    b. The local oscillator frequency
    c. The 30-MHz IF
    d. The pulse repetition frequency

13. The error voltage from the discriminator is applied to:
    a. The repeller (reflector) of the klystron
    b. The grids of the IF amplifier
    c. The grids of the RF amplifiers
    d. The magnetron

14. How do you eliminate stationary objects such as trees, buildings, bridges, etc., from the PPI presentation?
    a. Remove the discriminator from the unit
    b. Use a discriminator as a second detector
    c. Calibrate the IF circuit
    d. Calibrate the local oscillator

15. How far from the waveguide are the spark gap tubes?
    a. One quarter wavelength
    b. One half wavelength
    c. One wavelength
    d. Two wavelengths

16. If the radar unit has a PRR of 2000 Hz and a pulse width of 0.05 $\mu$s, what is the duty cycle?
    a. 0.0005
    b. 0.0001
    c. 0.05
    d. 0.001

17. The echo box is used for:

    a. Testing and tuning of the radar unit by providing artificial targets
    b. Testing the wavelength of the incoming echo signals
    c. Amplification of the echo signal
    d. Detection of the echo pulses

18. If the operating radar frequency is 3000 MHz, what is the distance between the waveguide and the spark gaps?

    a. 10 cm
    b. 5 cm
    c. 2.5 cm
    d. 20 cm

19. The heading flash is a momentary intensification of the sweep line on the PPI presentation. Its function is:

    a. To alert the operator when a target is within range
    b. To alert the operator when shallow water is near
    c. To inform the operator of the dead-ahead position on the PPI
    d. To inform the operator when the antenna is pointed to the rear of the ship

20. On what frequency is radar expected to cause interference?

    a. On the pulse repetition frequency
    b. On the klystron oscillator frequency
    c. On the magnetron frequency
    d. On most any communications frequency

21. How does radar interference appear on a loran scope?

    a. As stationary, narrow vertical spikes
    b. As narrow, vertical spikes moving across the screen
    c. As horizontal blanking fines
    d. As snow on the loran scope

22. The usual intermediate frequency of radar units is:

    a. 455 kHz
    b. 10.7 MHz
    c. 30 or 60 MHz
    d. 120 MHz

23. In a radar unit, the local oscillator is:

    a. A hydrogen thyratron
    b. A klystron
    c. A pentagrid converter tube
    d. A reactance tube modulator

24. How may the frequency of the klystron be varied?
    a. Small changes can be made by adjusting the repeller voltage
    b. Large changes can be made by adjusting the size of the resonant cavity
    c. By changing the phasing of the buncher grids
    d. Both a and b are correct

25. The magnetron is:
    a. A type of diode that requires an internal magnetic field
    b. A triode that requires an external magnetic field
    c. A type of diode that requires an external magnetic field
    d. Used as the local oscillator in the radar unit

26. If the magnetron is allowed to operate without the magnetic field in place:
    a. Its output will be somewhat distorted
    b. It will quickly destroy itself from excessive current flow.
    c. Its frequency will change slightly.
    d. Nothing serious will happen.

27. The radar serviceperson should take the following precautions to ensure that the magnet of the magnetron is not weakened:
    a. Keep metal tools away from the magnet.
    b. Do not subject it to excessive heat.
    c. Do not subject it to shocks or blows.
    d. All of the above.

28. What determines the minimum range of a radar set?
    a. The duty cycle
    b. The peak power output
    c. The time between the transmitter pulses
    d. The sensitivity of the discriminator.

29. What type of video output tube is used in radar units?
    a. A standard CRT
    b. A plan position indicator (PPI)
    c. An oscilloscope
    d. None of the above

30. What radar circuit determines the pulse repetition rate (PRR)?
    a. The discriminator
    b. The timer (synchronizer circuit)
    c. The artificial transmission line
    d. The pulse-rate-indicator circuit

31. If the PRR is 2000 Hz, what is the pulse repetition time?

    a. 0.05 s
    b. 0.005 s
    c. 0.0005 s
    d. 0.00005 s

32. To install and maintain a radar unit, you must have:

    a. Permission of the master but no license
    b. A first class radiotelephone or telegraph license with a radar endorsement
    c. An amateur extra class license
    d. A general radiotelephone license only

33. Who may replace fuses and receiving tubes in a radar unit?

    a. An unlicensed person
    b. A holder of a first class phone license with radar endorsement
    c. A holder of a general phone license with radar endorsement
    d. Any of the above

34. Under what circumstances can an unlicensed person operate a radar set?

    a. Only if the transmitter is a nontuneable type
    b. When the master of the ship designates him to operate it.
    c. Never
    d. Both a and b are correct

35. Range markers are determined by:

    a. The CRT
    b. The magnetron
    c. The timer
    d. The video amplifier

36. How are ranges changed so that range markers represent different distances?

    a. Change electrical potential on video amplifier
    b. Change the PPI screen
    c. Change the LC circuit, therefore, the oscillating frequency of the ringing oscillator
    d. Change the peak power

37. While making repairs or adjustments to radar unit:

    a. Wear gloves
    b. Discharge all high-voltage capacitors to ground
    c. Maintain filament voltage
    d. Reduce magnetron voltage

38. While working with a CRT, it is a good idea to:

    a. Discharge the first anode
    b. Test the second anode with your fingertip
    c. Wear gloves and goggles
    d. Set it down on a hard surface

39. Sea return is:

    a. Sea water that gets into the antenna system
    b. The return echo from a target at sea
    c. Reflection of radar signal from nearby waves
    d. None of the above

40. What is the distance in nautical miles to a target if it takes 310 $\mu$s for the radar pulse to travel from the radar antenna to the target and back.

    a. 12.5 mi
    b. 25 mi
    c. 50 mi
    d. 2.5 mi

41. The sensitivity-time control (STC) circuit:

    a. Increases the sensitivity of the receiver for close objects
    b. Increases the sensitivity of the receiver for distant objects
    c. Decreases the sensitivity of the receiver for close objects
    d. Decreases the sensitivity of the transmitter for close objects

42. The primary tube used in the STC circuit is:

    a. A pentode
    b. A triode
    c. A diode
    d. A hydrogen thyratron

43. In a radar unit, the mixer uses:

    a. A silicon crystal
    b. A pentagrid converter tube
    c. A field-effect transistor
    d. A microwave transistor

44. Why is hydrogen gas used in thyratron tubes?

    a. It ionizes and deionizes slowly.
    b. It ionizes and deionizes quickly.
    c. Because it is much lighter than other gases.
    d. Because it does not ionize at all.

45. The timer circuit:

    a. Determines the PRR
    b. Determines range markers
    c. Provides blanking and unblanking signals for the CRT
    d. All of the above are correct

46. What component is block 46 in the block diagram of Fig. 22-29?

    a. ATR box
    b. TR box
    c. RF attenuator
    d. Crystal detector

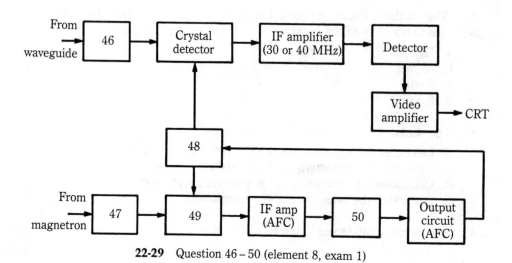

**22-29**   Question 46 – 50 (element 8, exam 1)

47. What component is block 47 in the block diagram of Fig. 22-29?

    a. ATR box
    b. TR box
    c. RF attenuator
    d. Crystal detector

48. What component is block 48 in the block diagram of Fig. 22-29?

    a. Discriminator
    b. IF amplifier
    c. Klystron
    d. Crystal detector

49. What component is block 49 in the block diagram of Fig. 22-29?

    a. Discriminator
    b. IF amplifier

    c. Klystron

    d. Crystal detector

50. What component is block 50 in the block diagram of Fig. 22-29?

    a. Discriminator

    b. IF amplifier

    c. AFC amplifier

    d. Crystal detector

# Element 8, exam 2

1. Silicon crystals are used in radar mixer and detector stages. Using an ohm-meter, how might a crystal be checked to determine if it is functional?

    a. Its resistance should be the same in both directions.

    b. Its resistance should be low in one direction and high in the opposite direction.

    c. Its resistance cannot be checked with a dc ohmmeter because the crystal acts as a rectifier.

    d. It would be more appropriate to use a VTVM and measure the voltage drop across the crystal.

2. Silicon crystals:

    a. Are very sensitive to static electric charges

    b. Should be wrapped in lead foil for storage

    c. Tolerate very low currents

    d. All of the above are correct

3. The TR box:

    a. Prevents the received signal from entering the transmitter

    b. Protects the receiver from the strong radar pulses

    c. Turns off the receiver when the transmitter is on

    d. Both b and c are correct

4. The dc keep-alive potential:

    a Is applied to the TR tube to make it more sensitive.

    b. Partially ionizes the gas, making it very sensitive to the transmitter pulse

    c. Fully ionizes the gas

    d. Both a and b are correct

5. If the crystal mixer becomes defective, replace the:

    a. Crystal only

    b. The crystal and the ATR tube

    c. The crystal and the TR tube

    d. The crystal and the klystron

6. The TWT (traveling wave tube) is:

   a. A type of waveguide
   b. Not really a vacuum tube but a semiconductor device
   c. A microwave amplifier tube
   d. A microwave tube that requires a very strong external magnetic field for proper operation

7. The input signal to a TWT is inserted:

   a. At the anode end of the helix
   b. At the grid of the tube
   c. At the cathode
   d. At the cathode end of the helix

8. Unblanking pulses are produced by the timer circuit. Where are they sent?

   a. To the IF amplifiers
   b. To the CRT
   c. To the mixer
   d. To the discriminator

9. At microwave frequencies, waveguides are used instead of conventional coaxial transmission lines because:

   a. They are smaller and easier to handle.
   b. They have considerably less loss.
   c. They are lighter since they have hollow centers.
   d. Moisture is never a problem with them.

10. Long horizontal sections of waveguides are not desirable because:

    a. The waveguide can sag, causing loss of signal.
    b. Moisture can accumulate in the waveguide.
    c. Excessive standing waves can occur.
    d. The polarization of the signal might shift.

11. How long would it take for a radar pulse to travel to a target 10 nautical miles away and return to the radar receiver?

    a. 124 $\mu$s
    b. 12.4 $\mu$s
    c. 1.24 $\mu$s
    d. 10 $\mu$s

12. What is the average power if the radar set has a pulse repetition rate of 1000 Hz, a pulse width of 1 $\mu$s, and a peak power rating of 100 kW?

    a. 10 W
    b. 100 W

c. 1000 W

d. None of the above

13. How long is a half wavelength at 5400 MHz?

    a. 5.5 cm
    b. 2.7 cm
    c. 11 cm
    d. 55 cm

14. What determines the minimum range of a radar set?

    a. The duty cycle
    b. The average power
    c. The peak power
    d. The time between transmitted pulses

15. Which is a typical current for a silicon crystal used in a radar mixer or detector circuit?

    a. 3 mA
    b. 15 mA
    c. 50 mA
    d. 100 mA

16. What does the following schematic in Fig. 22-30 represent?

    a. A magnetron circuit
    b. A klystron oscillator
    c. An STC circuit
    d. An audio oscillator

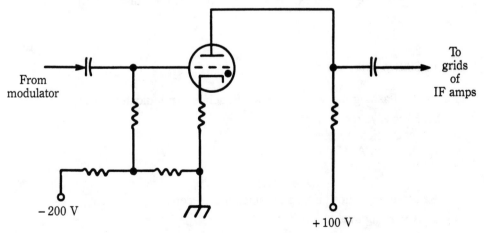

**22-30**  Question 16 (element 8, exam 2)

The next two questions are based on the diagram in Fig. 22-31.

17. At the operating frequency of 3000 MHz, what is the distance between the waveguide and the ATR spark gap?

    a. 10 cm
    b. 5 cm
    c. 2.5 cm
    d. 1.25 cm

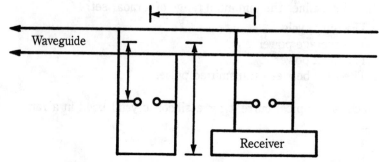

**22-31**  Question 17 and 18 (element 3, exam 2)

18. At the operating frequency of 3000 MHz, what is the distance between the waveguide and the receiver?

    a. 10 cm
    b. 5 cm
    c. 2.5 cm
    d. 1.25 cm

19. If the pulse-repetition rate is 2500 Hz, what is the pulse-repetition time?

    a. 40 $\mu$s
    b. 400 $\mu$s
    c. 250 $\mu$s
    d. 800 $\mu$s

20. Bright flashing pie sections on a PPI scope can indicate:

    a. Excessive receiver gain
    b. Insufficient receiver gain
    c. Defect in AFC circuit
    d. Interference

21. Radar interference on a communications receiver appears as:

    a. A varying tone
    b. Static
    c. A steady tone
    d. A hissing tone

22. Who is permitted to operate a ship radar unit?
    a. Only FCC-licensed persons with ship radar endorsement
    b. Only the engineer or someone under his or her direct supervision
    c. The master of the ship, or anyone designated by the master
    d. Anyone who is knowledgeable

23. Oscillations in a klystron local oscillator tube are maintained:
    a. By the grid-feedback loop
    b. By bunches of electrons passing the cavity grids
    c. By the circulation of electrons
    d. By the LC circuit

24. The STC circuit:
    a. Increases the sensitivity of the receiver for close targets
    b. Decreases sea return
    c. Helps to increase the bearing resolution
    d. Increases sea return

25. The klystron local oscillator is constantly kept on frequency by:
    a. Constant manual adjustments
    b. The AFC circuit
    c. A feedback loop from the crystal detector
    d. A feedback loop from the TR box

26. In a radar receiver, the RF amplifier:
    a. Is high gain
    b. Is low gain
    c. Requires a wide bandwidth
    d. Does not exist

27. Fine adjustment of a reflex klystron is accomplished by:
    a. Adjusting the flexible wall of the cavity
    b. Varying the repeller voltage
    c. Adjusting the AFC control system
    d. Varying the cavity grid potential

28. The pulse-repetition rate (PRR) refers to:
    a. The reciprocal of the duty cycle
    b. The pulse rate of the local oscillator tube
    c. The pulse rate of the magnetron
    d. The pulse rate of the klystron

29. How is the signal removed from a waveguide or magnetron?
    a. With a T-hook
    b. With a J-Hook

   c. With a coaxial connector
   d. With a waveguide flange joint

30. What is the average range of a ship radar unit?
   a. 20 mi
   b. 40 mi
   c. 80 mi
   d. 100 mi

31. Before testing a radar transmitter, it would be a good idea to:
   a. Make sure no one is on the deck
   b. Make sure magnetron magnetic field is far away from the magnetron
   c. Make sure there are no explosives or inflammable cargo being loaded
   d. Make sure the master has been notified

32. How is radar interference to a communications receiver eliminated?
   a. By not operating other devices when radar is in use
   b. By properly grounding, bonding, and shielding all units
   c. By using a highpass filter on the power line
   d. By using link coupling

33. A defective crystal in the AFC section will cause:
   a. No serious problem
   b. Bright flashing pie sections on the PPI
   c. Spiking on the PPI
   d. Vertical spikes that constantly move across the screen

34. A small hole is sometimes drilled in the lowest point of a waveguide:
   a. Never, because it will affect the resonant frequency
   b. To make fine adjustments in the resonant frequency
   c. To allow condensed moisture to drain out
   d. To help equalize the temperature

35. The anode of a magnetron is normally maintained at ground potential:
   a. Because it operates more efficiently that way
   b. For safety purposes
   c. Never. It must be highly positive to attract the electrons.
   d. Because greater peak-power ratings can be achieved

36. The keep-alive voltage is applied to:
   a. The crystal detector
   b. The ATR tube
   c. The TR tube
   d. The magnetron

37. The purpose of the discriminator circuit in a radar set is:

    a. To discriminate against nearby objects
    b. To discriminate against two objects with very similar bearings
    c. To generate a corrective voltage for controlling the frequency of the klystron local oscillator
    d. To demodulate or remove the intelligence from the FM signal

38. How can a CRT be damaged?

    a. By operating it at a high intensity
    b. By operating it at a lower-than-normal intensity
    c. By leaving a stationary high-intensity image on the screen
    d. By operating it beyond its frequency limit

39. If a CRT is dropped:

    a. Most likely nothing will happen because they are built with durability in mind.
    b. It might go out of calibration.
    c. It might implode, causing damage to workers and equipment.
    d. The phosphor might break loose.

40. When a radar signal is sent to an object, the Doppler effect is:

    a. Where objects moving toward you reflect back a lower frequency
    b. Where objects moving away from you reflect back a higher frequency
    c. Where stationary objects reflect back a slightly lower frequency
    d. Where objects moving toward you reflect back a higher frequency

41. What maintenance work may be done by unlicensed workers?

    a. None
    b. Replacement of magnetron and klystron tubes
    c. Replacement of fuses and receiving-type tubes only
    d. Minor frequency adjustments only

42. A high magnetron current indicates:

    a. Defective AFC crystal
    b. Defective external magnetic field
    c. An increase in duty cycle
    d. A high standing-wave ratio

43. What does a loran scope indicate when it experiences interference from a radar transmitter?

    a. Vertical spikes moving across the screen
    b. Spikes on the screen
    c. Bright pie sections on the screen
    d. Grass on the screen
    e. Both a and d are correct

44. If the TR box malfunctions:

    a. The transmitter might be damaged
    b. The receiver might be damaged
    c. The klystron might be damaged
    d. Magnetron current will increase

45. What is the cavity between the magnetron and the mixer?

    a. TR
    b. ATR
    c. Attenuator
    d. Resonant cavity

46. What should be done to the interior surface of a waveguide in order to minimize signal loss?

    a. Fill with nitrogen gas
    b. Paint it with nonconductive paint to prevent rust
    c. Keep it as clean as possible
    d. Fill with a high-grade oil

47. How do you prolong the life of a spark gap?

    a. Apply only low voltages
    b. Periodically reverse the polarity
    c. Apply extremely high voltages from time to time
    d. It does not matter because spark gaps are not used anymore

48. If long-length transmission lines are not properly shielded and terminated:

    a. Silicon crystal can be damaged.
    b. Communications receiver interference might result.
    c. Overmodulation might result.
    d. Excessive RF loss can result.

49. The following is true concerning waveguides:

    a. Conduction is accomplished by polarization of electromagnetic and electrostatic fields.
    b. The magnetic field is strongest at the edges of the waveguide.
    c. The skin effect is employed.
    d. Both a and b are correct.

50. How does a TWT amplify?

    a. Through a positive feedback circuit
    b. Through a negative feedback circuit
    c. Like any other triode tube
    d. By transfer of energy from the signal to the electron beam

# Appendix A
# Answer key

## Chapter 1
## Rules and regulations

1. 3 days.

2. master of vessel, or
   owner of vessel, or
   ship station licensee

3. at least 60 days before it is needed

4. a portable VHF hand-held unit used for communication with other associated ship units and with the ship it is associated with

5. the master of the vessel

6. to attract attention of persons on watch or to activate automatic alarm devices

7. that a distress call is about to follow, or
   that an urgent cyclone warning is about to follow, or
   that persons have fallen overboard

8. 8 kHz

9. any frequency that contains 0.25% of the total radiated power

10. navigational purposes only

11. in an emergency situation, or
    when rounding the bend of a river or navigating through a blind spot, or
    when a ship fails to respond to a call on low power

12. channel 13 (156.650 MHz)

13. a frequency for making brief calls, in order to make initial contact. Once
contact is made, the stations must move to a working frequency

14. no

15. Modification of license applies whenever any of the following are affected
    by the change:
    a. frequency tolerance, or
    b. modulation, or
    c. emission, or
    d. power, or
    e. bandwidth

16. a ship that is required to have radio communications equipment

17. signature of licensed operators,
    list of frequencies used,
    emission types,
    starting and ending times of communications;
    See *Contents of station records* and *logs* for additional details.

18. details of what the technician or anyone under the technician's supervision,
has done

19. the operating position of the station

20. Listen before transmitting.
    Keep transmission short.
    Do not engage in superfluous communications.
    Do not transmit general calls not addressed to a particular station.
    Do not transmit while on land.
    Do not use selective calling on 2182 kHz or 156.8 MHz. (Tone-coded sig-
    nals designed to deactivate the squelch of only specific stations.)

21. SOS

22. mayday

23. the authority of the master

24. 2182 kHz

25. 500 kHz

26. those who are U.S. citizens, or
    an alien who is eligible for employment in the U.S., or
    those who hold a valid U.S. pilot certificate

27. if you are deaf or mute, or
    if you are unable to speak English, or
    if your license is under suspension, or
    if you are involved in any pending litigation based on alleged violations.

28. The ELT is an emergency locator transmitter. They are tested with a dummy load and in coordination with the FAA. ELTs are used for aviation purposes while EPIRBs are used for marine purposes.

29. to facilitate search and rescue operations

30. following use in an emergency situation, or
    before the expiration date, established by the manufacturer (when 50% of its useful life has been expired)

31. J3E = single-sideband suppressed carrier amplitude modulated telephony
    H3E = single-sideband full carrier amplitude modulated telephony
    R3E = single-sideband reduced carrier amplitude modulated telephony
    A3E = double-sided full carrier amplitude modulated telephony

32. ±5 kHz

33. Airborne transmitter frequency measurements are as required. Nonairborne transmitter frequency measurements must be done:
    a. when transmitter is initially installed, and
    b. when any change or adjustment is made in the transmitter that affects its transmitting frequency, and
    c. whenever there is reason to believe that an operating frequency has shifted beyond tolerance, and
    d. upon receipt of an official notice of off-frequency operation.

34. 20 Hz below 156 MHz

35. See Tables 1-3 and 1-4.

36. the authority to install, adjust, maintain, and operate marine or aviation radio equipment.

37. radio operator must facilitate the inspection with cooperation.

38. 5 years

39. Listening watch is continuous on 2182 kHz whenever the station is not transmitting.

40. routine logs—one year
    logs involving distress traffic—three years
    logs that include communications under FCC investigation—maintained until FCC gives written notice that they can be destroyed

41. summary of distress traffic
    daily position of ship
    results of equipment tests
    maintenance details
    Please see logs for additional details.

42. Apply for a new license.

43. A 5-year permit that authorizes its holder to operate only those transmitters that have simple, external switching devices. Transmitters must not require adjustment of frequency-determining elements.

44. between 75% and 100%

45. name of manufacturer and type or model number

46. while the ship is being transported, stored, or parked on land

47. Listen first to assure a clear frequency.
    Initiate the call on a calling frequency.
    Move to a working frequency for the communication.

48. *Over* means you are finished transmitting, but expect a response.
    *Clear* means you are finished transmitting, and you do not expect a response.
    *Roger* means that you received the transmission correctly.

49. unnecessary transmissions
    unidentified transmissions
    willful or malicious interference

signals not addressed to a particular station
Please see *operating procedure* for more examples.

50. when communications are difficult because of weak signals or noise.

51. with other portable ship units
with the mother ship

52. 1 W

53. give the call sign of the vessel you are associated with, followed by appropriate unit designator.

54. no

55. at the principal operating position of the station.

56. distress signals, urgency signals, and safety signals.

57. urgency signal

58. safety signal

59. with no one

60. Only emergency communications can be transmitted during the silent period. Silent period is for three minutes in duration twice per hour (on the hour and on the half hour).

61. distress, urgency, and vital navigational warnings.

62. ship station assigned call sign spoken in English. Bridge-to-bridge stations may use name of vessel in lieu of call sign.

63. if transmitter is off frequency, or
if signal becomes distorted, or
if modulation percent exceeds 100%

64. Announce call sign, followed by the word test.
If another station says *wait*, suspend the test for 30 seconds.
Test must not exceed 10 s.
Finish with call sign.

65. daily

66. General Radiotelephone Operator License

67. 150 mi

68. type approval = FCC conducts equipment tests to assure specifications
type acceptance = manufacturer conducts equipment tests to assure specifications

69. 10 days

70. Fines range from $1,000 to $25,000 and up to 1 year imprisonment, See Table 1-7 for details.

71. 40 dB below peak envelope power

72. National Bureau of Standards

73. post original license at one location and carry a photocopy to other locations

74. 156.800 MHz

75. Channel 22 on the VHF band and 2182 kHz on the medium frequency band.

76. A mode of operation in which two stations can talk at the same time.

77. ELTs operate on 121.5 MHz and 243.0 MHz.

78. Class A or class B EPIRB must be on board.

79. Class C

80. Power must not be less than 75 mW after 48 h of continuous operation.

81. 50 Hz

82. The General Radiotelephone Operator License and The Marine Radio Operator Permit.

83. A frequency the radio operator is expected to move to, once the initial contact is made on the calling frequency.

84. No. An aircraft license cannot be assigned or transferred. The new owner must file for a new license. (Rules and Regulations 87.33)

85. Only if they have been granted Temporary Operating Authority by the FCC. Then they may operate for up to 90 days.

86. Special temporary authorization may be granted for a period not to exceed 180 days. However, this appears to be very difficult to acquire. Several statements must be made regarding: services to be rendered; types of emissions to be used; a statement of noninterference; etc.

# Chapter 2
# Glossary

1. to check tank circuit frequency
   to check field strength of an antenna
   to check output frequency of a transmitter

2. it should be kept as far as possible from the oscillator being measured.

3. effective

4. current is maximum and voltage is minimum
   current is minimum and voltage is maximum

5. longer, inductor

6. detector stage

7. RF and IF amplifiers

8. decreased

9. total radiated power

10. hydrometer

11. distilled water

12. steady, light

13. parallel

14. series

15. resistance

16. lower; less stable

17. indefinitely

18. insulation; RF energy

19. sharp; refraction

20. emitter

21. low, high

22. rectifier

23. pre-amp

24. loose

25. parallel. (When capacitance in a tank circuit is increased, the resonant frequency is lowered. Placing a capacitor in parallel increases capacitance.)

26. series. (A capacitor in series lowers the capacitance, which increases the resonant frequency. See *Tank Circuits* in *Transmitters* chapter.)

27. temperature

28. insufficient

29. capacitance, resistance

30. power supply regulation

31. permanent

32. step up

33. by controlling dc input with a rheostat

34. iron core

35. C

36. A

37. A

38. fixed

39. high

40. lengthens the life of the filament

41. distortion

42. C

43. frequency multiplier

44. resonant frequency

45. ground

46. is oscillating

47. neutralization

48. 700 Hz

49. zero beat

50. maximum power

51. square; number of turns

52. input impedance; output impedance

53. intermediate

54. image reception

55. electrical energy, watt-hour

56. as short as possible

57. low, constant

58. two widely separated circuits

59. zero

60. modulation index

61. neutralization

62. B+

63. moisture

64. capacitance; inductance

65. ohmmeter

66. omnidirectional

67. directivity; unidirectionality

68. peak inverse voltage

69. 75; 100

70. current; voltage

71. voltage; current

72. flywheel

73. average

74. even

75. capacitance; inductance

76. no-load, full-load

77. removed

78. less

79. one third

80. inductive reactance; capacitive reactance

81. insulation

82. loosely, strong

83. attenuate adjacent signals

84. loose

85. load

86. one sixth

87. filtered out; balanced out

88. surface

89. odd

90. directivity

91. impedance

92. 63.2

93. forward; reverse

94. less

95. collector

96. resistance; current

97. cathode

98. electrical conduction

99. capacitance; voltage

100. high, load

101. voltage, current, phase differences

102. true

103. electrical energy

104 frequency

105. size

106. Amplitude modulation (AM) and frequency modulation (FM)

107. The horizontal position extending 360° around the horizon

108. Because they offer minimal signal loss across the mechanical joint.

109. vertical

110. increase

111. From the top view, number them counter-clockwise around the IC, starting from the notch.

112. 110 kHz

113. 6080 feet or 1.1516 satute mile

114. 12.4

115. Wetting is the solderability of the materials. Good wetting means the solder flows freely like a fluid.

116. a dirty surface
     insufficient heat

117. 0200 hours the next day

118. 3 – 30 kHz

# Chapter 3
# Emission types and frequency ranges

1. 30 MHz

2. medium frequency band

3. A3E (formerly A3)

4. Single sideband suppressed carrier (SSSC). The brief designation is J3E.

5. C3F

6. Facsimile, F3C

7. in the station log book

8. H3E

9. 3 – 30 kHz

10. 6M25C3F

11. 100HA1A

# Chapter 4
# Direct current

1. drift, difference of potential

2. EMF
   Difference of potential
   *IR* drop

3. ohms per volt

4. ohmmeter

5. cross-sectional area

6. one third

7. decreases

8. series

9. free electrons

10. mica
    quartz
    Teflon
    polystyrene

11. $P = I^2R$

12. hydrometer

13. drain electrolyte and fill the battery with distilled water

14. ampere-hours

15. about 15 W

16. resistance must be four times its original value.

17. The number of electrons in the valence shell

18. 25 Ω

19. 230 KΩ with 5% tolerance

20. 2 hours

21. 20 V

# Chapter 5
# Alternating current

1. period = $\dfrac{1}{\text{frequency}}$

2. 360

3. rms (effective)

4. 33.3 $\mu$F

5. 600 $\mu$F

6. decreases

7. output would steadily increase

8. series

9. 12.5 $\mu$H

10. zero

11. current; voltage; 90

12. 45°

13. 55°

14. 0°

15. 63.2; voltage

16  0.02 s
　　5
　　0.1 s

17. The inductance would increase.

18. 1.414 MΩ

19. 0.002182 GHz

20. The capacitor charges immediately

21. Since the capacitor is connected directly across the battery, it charges instantly. See Fig. 5-8c.

22. At one time constant, when it has discharged down 63.2%.

# Chapter 6
# Transformers

1. step-up

2. $\dfrac{N_p}{N_s} = \dfrac{E_p}{E_s} = \dfrac{I_s}{I_p}$

where
　　$I$ = current
　　$E$ = Voltage
　　$N$ = Number of turns
　　$p$ = Primary
　　$s$ = Secondary

3. 660 V

4. 13.2 A

5. $3.3 \times 10^{-3}$A or 0.0033 A

6. 25:1

7. Eddy currents

8. Copper

# Chapter 7
# Semiconductors

1. low; forward

2. high; reverse

3. p; n

4. cathode

5. grid

6. plate

7. forward

8. reverse

9. less

10. low, high

11. common emitter

12. collector current increases

13. See text.

14. very high

15. voltage regulation

16. varactor

17. external

18. $\text{Gain} = -\dfrac{\text{Feedback resistance}}{\text{Input resistance}}$

19. 5 V

# Chapter 8
# Power supplies

1. twice

2. lower

3. excessive PIV

4. voltage transients

5. 12

6. changing load conditions

7. 1 A

# Chapter 9
# AM and FM receivers

1. C

2. incoming signal frequency

3. incoming signal
   local oscillator signal
   their sum
   their difference

4. demodulate

5. second detector; RF and IF amplifiers

6. loose

7. low

8. split tuning; looser

9. attenuate

10. limiter

11. discriminator

# Chapter 10
# Transmitters

1. buffer amplifier

2. modulator

3. LC circuit

4. quartz

5. temperature variations

6. thickness

7. series

8. fidelity; efficiency

9. frequency multipliers

10. class C

11. flywheel

12. multiple

13. 15,274 kHz (15.274 MHz)

14. one fourth

15. third

16. 100

17. one third

18. See text.

19. 500 W

20. 25 W

21. tuning and loading

22. tuning; loading

23. any circuit

24. parasitic

25. a. shielding
    b. neutralization
    c. RF chokes
    d. noninductive resistors

26. increases

27. B+

28. a. neon bulb
    b. grid-dip meter
    c. thermocouple ammeter
    d. flashlight bulb and wire loop

29. zero

30. RF amps, grounded grid

31. transmitter; receiver

32. low, high

# Chapter 11
# Bandwidth of emission

1. bandwidth of emission

2. carrier, sidebands; harmonics

3. a. carrier
   b. upper sideband
   c. lower sideband

4. a. incoming frequency
   b. local oscillator
   c. sum of the above
   d. difference of the above

5. a. 4 kHz
   b. 2 kHz

6. a. disregard the carrier
   b. multiply the modulating tone by 2.

7. carrier, sidebands

8. a. disregard the carrier.
   b. $\# \times (F) \times (2)$ where $\#$ is the number of significant sidebands.

9. 180 kHz

10. 50 kHz

# Chapter 12
# Antennas

1. resonant frequency

2. shorter

3. decrease

4. decreases

5. 3 m

6. 60 cm

7. 10 MHz

8. Hertz

9. horizontally

10. See text.

11. See text.

12. Current is maximum; voltage is minimum.

13. Current is minimum; voltage is maximum.

14. vertically

15. grounded

16. series

17. shunt; zero

18. adding parasitic elements; stacking antennas

19. directivity; improved

20. increase; more

# Chapter 13
# Transmission lines

1. center conductor; shield

2. nitrogen; moisture

3. impedance mismatch

4. standing, remain constant

5. mismatch

6. power loss

7. standing waves; infinite

8. See Fig. 13-4.

9. inductive reactance

10. Tables 13-1 and 13-2.

11. 50 Ω

# Chapter 14
# Effective radiated power

1. The transmitter power must be 16 times stronger.

2. 75 $\mu$V/m.

3. 708 W of ERP

4. The loss increases.

5. 1.26, 2, and 10.

# Chapter 15
# UHF and above

1. decreases

2. short

3. capacitance; inductance

4. copper foil

5. surface

6. waveguides

7. waveguides

8. transit

9. traveling wave tube

10. cathode

# Chapter 16
# Motors and generators

1. mechanical rectifier

2. dirty brushes

3. a. dirty commutator
   b. worn-out brushes
   c. brushes in neutral position
   d. excessive play

4. sparking

5. the wire insulation

6. residual magnetism

7. rotate; 90°

8. destroy itself (if the motor has no load)

9. the load

10. step up dc voltage

11. a rheostat in series with the battery source

12. motor-generator

13. rugged, high

14. less

15. motor-generator

16. See Table 16-2.

# Chapter 17
# Measurements

1. Power $= \dfrac{E^2}{R}$

2. See Fig. 17-5.

3. true power

4. voltage; current; phase differences

5. watthour meter

6. ohmmeter

7. ohmmeter

8. To check tank-circuit frequency.
   To check the output frequency of an antenna.
   To check the output frequency of a transmitter.

9. The wavemeter should be kept as far away as possible from the circuit being measured.

10. resonant frequency

11. 196 Hz

12. 10 dB.

13. The potential on the grid.

14. Shields the electron beam from external electric fields. Produces a final acceleration of the electron beam. Attracts the secondary emission electrons from the screen.

# Chapter 18
# Digital circuitry

1. A high on input E will make output low.

2. A high on input E will make output high.

3. A high on input E will make output

| | | | |
|---|---|---|---|
| a. 1 | f. 1 | j. 0 | n. 1 |
| b. 0 | g. 1 | k. 0 | o. 0 |
| c. 0 | h. 1 | l. 1 | p. 0 |
| d. 0 | i. 0 | m. 0 | q. 0 |
| e. 1 | | | |

# Chapter 19
# Avionics

1. very low frequency range

2. 328 yards one way

3. 108 MHz – 118 MHz

4. Provides vertical positioning information

5. Provides horizontal positioning information

# Chapter 22
## Element 3, exam 1

| | | | |
|---|---|---|---|
| 1. e | 26. d | 51. d | 76. d |
| 2. e | 27. d | 52. d | 77. d |
| 3. b | 28. c | 53. b | 78. c |
| 4. d | 29. d | 54. e | 79. d |
| 5. b | 30. c | 55. b | 80. d |
| 6. b | 31. a | 56. c | 81. e |
| 7. e | 32. d | 57. d | 82. c |
| 8. c | 33. d | 58. d | 83. c |
| 9. a | 34. c | 59. a | 84. e |
| 10. c | 35. d | 60. e | 85. e |
| 11. b | 36. e | 61. c | 86. d |
| 12. b | 37. c | 62. b | 87. b |
| 13. d | 38. d | 63. c | 88. e |
| 14. e | 39. c | 64. c | 89. e |
| 15. a | 40. b | 65. e | 90. d |
| 16. b | 41. d | 66. e | 91. d |
| 17. c | 42. d | 67. d | 92. b |
| 18. e | 43. c | 68. e | 93. c |
| 19. b | 44. d | 69. e | 94. b |
| 20. e | 45. c | 70. c | 95. a |
| 21. c | 46. d | 71. c | 96. d |
| 22. e | 47. b | 72. c | 97. b |
| 23. d | 48. e | 73. d | 98. d |
| 24. d | 49. d | 74. a | 99. c |
| 25. c | 50. d | 75. d | 100. c |

## Element 3, exam 2

| | | | |
|---|---|---|---|
| 1. e | 14. c | 27. e | 40. e |
| 2. d | 15. a | 28. d | 41. d |
| 3. c | 16. e | 29. c | 42. b |
| 4. d | 17. e | 30. b | 43. b |
| 5. d | 18. e | 31. a | 44. d |
| 6. c | 19. c | 32. b | 45. e |
| 7. d | 20. e | 33. d | 46. e |
| 8. a | 21. c | 34. a | 47. e |
| 9. e | 22. d | 35. c | 48. d |
| 10. c | 23. a | 36. a | 49. e |
| 11. d | 24. e | 37. d | 50. a |
| 12. b | 25. e | 38. e | 51. c |
| 13. d | 26. e | 39. d | 52. d |

| | | | |
|---|---|---|---|
| 53. b | 65. c | 77. a | 89. b |
| 54. c | 66. b | 78. b | 90. b |
| 55. c | 67. b | 79. c | 91. c |
| 56. d | 68. c | 80. a | 92. d |
| 57. c | 69. c | 81. a | 93. c |
| 58. b | 70. c | 82. c | 94. d |
| 59. c | 71. a | 83. b | 95. b |
| 60. c | 72. b | 84. b | 96. d |
| 61. c | 73. c | 85. b | 97. c |
| 62. b | 74. c | 86. a | 98. c |
| 63. b | 75. c | 87. d | 99. e |
| 64. b | 76. c | 88. b | 100. b |

## 1 Element 8, exam 1

| | | | |
|---|---|---|---|
| 1. b | 14. b | 27. d | 39. c |
| 2. d | 15. a | 28. c | 40. b |
| 3. c | 16. b | 29. b | 41. c |
| 4. e | 17. a | 30. b | 42. d |
| 5. c | 18. c | 31. c | 43. a |
| 6. c | 19. c | 32. b | 44. b |
| 7. b | 20. f | 33. d | 45. d |
| 8. c | 21. b | 34. d | 46. b |
| 9. b | 22. c | 35. c | 47. c |
| 10. c | 23. b | 36. c | 48. c |
| 11. a | 24. d | 37. b | 49. d |
| 12. c | 25. c | 38. c | 50. a |
| 13. a | 26. b | | |

## 2 Element 1, exam 1

| | | | |
|---|---|---|---|
| 1. b | 14. d | 27. b | 39. c |
| 2. d | 15. a | 28. c | 40. d |
| 3. d | 16. c | 29. b | 41. c |
| 4. d | 17. c | 30. b | 42. b |
| 5. c | 18. b | 31. c | 43. e |
| 6. c | 19. b | 32. b | 44. b |
| 7. d | 20. c | 33. b | 45. b |
| 8. b | 21. c | 34. c | 46. c |
| 9. b | 22. c | 35. b | 47. b |
| 10. b | 23. b | 36. c | 48. b |
| 11. a | 24. b | 37. c | 49. b |
| 12. b | 25. b | 38. c | 50. d |
| 13. b | 26. d | | |

# Appendix B
# Troubleshooting, standard phonetic alphabet, and time zoners

## Troubleshooting

You can expect to see a diagram similar to Fig. B-1 on your FCC test. You might be asked several questions that relate to such a circuit. The diagram is taken from *The FCC Test Handbook* and used with the permission of WPT Publications.

**C1**  Can be used to decrease oscillation of the crystal. The XTAL (crystal) acts like a tank circuit. A capacitor connected in parallel increases the tank circuit capacitance because parallel capacitors add. Increased capacitance results in decreased resonant frequency. A capacitor connected in series would increase the crystal frequency.

**XTAL**  The thickness of the quartz crystal is the primary determiner of its oscillation frequency. Crystal frequency will drift with temperature changes. It is, therefore, important to maintain a constant crystal temperature.

**Q1**  This field-effect transistor is characterized by a very high impedance.

**C2**  The coupling capacitor couples the oscillator signal to Q2 (frequency multiplier). If it becomes open, only the oscillator will function. The signal will not make it to Q2.

**C3**  This is a power-supply bypass capacitor. Its purpose is to bypass any ac in the power supply to ground. If open, very little change would result. Of all the circuit components, removal of this capacitor would influence the circuit the least. Of course, if C3 were shorted, the power supply would be shorted to ground.

**C4**  This is the emitter bypass capacitor. If shorted, the emitter resistor will be shorted out. The transistor bias would change. This may distort the output. If C4 opens, degeneration occurs. This results in a reduced amplifier gain and, therefore, reduced output. The output will be more stable.

**C5**  This is the tank circuit tuning capacitor. As it is adjusted, the resonant frequency of the tank is changed. This enables you to tune to any harmonic of the

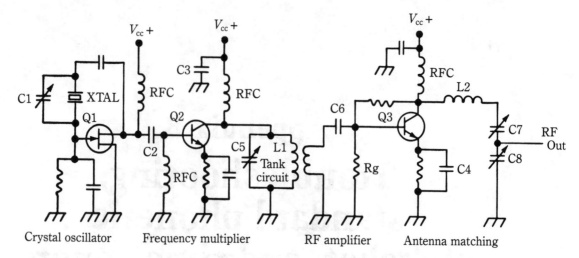

**B-1**   Troubleshooting.

input frequency. For example, if the input frequency is 12 MHz and you want to form a frequency doubler (tune output to 24 MHz), you must reduce the value of C5 to one quarter of its 12-MHz resonant value. This can be verified with the resonant frequency formula in the text.

**C6**   If this coupling capacitor becomes open, no signal will reach the RF amp. If "shorted," the base of Q3 would be shorted to ground, resulting in a change in the transistor bias.

**C7**   This portion of the antenna-matching network is the tuning capacitor. As you can see, it is a series resonant circuit. As C7 is adjusted, the resonant frequency changes. The tuning capacitor is tuned in such a way that it resonates at the operating frequency of the transmitter (final RF amplifier).

**C8**   This is the loading capacitor. By adjusting it, the output of the transmitter more closely matches the load. When the impedance of the transmitter is properly matched to the load (transmission line and antenna), maximum power is transferred from transmitter to load.

**Rg**   In a tube circuit, this is the grid resistor. In this case, it is the base resistor. It controls the bias, and therefore, the class of amplification of Q3.

# Standard phonetic alphabet

When conditions are poor or signals weak, the radio operator must resist the temptations of shouting into the microphone or turning up the microphone gain control. These will only overmodulate and distort your signal.

The squelch is usually adjusted to allow the receiver to be heard only when signals above the noise level are present. Very weak signals might be at the noise level. When they are this weak, they will not be heard unless you open up the squelch all the way (turn if off).

Another way to communicate accurately, when conditions are poor, is to use the Standard Phonetic Alphabet. Because everyone learns the same phonetics, words are easy to recognize. Table B-1 is summarized from FCC Bulletin FO-32:

**Table B-1. Phonetic alphabet.**

| Letter | Word | Pronunciation |
|--------|------|---------------|
| A | Alfa | *AL* FAH |
| B | Bravo | *BRAH* VOH |
| C | Charlie | *CHAR* LEE |
| D | Delta | *DELL* TAH |
| E | Echo | *ECK* OH |
| F | Foxtrot | *FOKS* TROT |
| G | Golf | *GOLF* |
| H | Hotel | HOH *TELL* |
| I | India | *IN* DEE AH |
| J | Juliett | *JOO* LEE *ETT* |
| K | Kilo | *KEY* LOH |
| L | Lima | *LEE* MAH |
| M | Mike | MIKE |
| N | November | NO *VEM* BER |
| O | Oscar | *OSS* CAH |
| P | Papa | PA *PAH* |
| Q | Quebec | KEH *BECK* |
| R | Romeo | *ROW* ME OH |
| S | Sierra | SEE *AIR* RAH |
| T | Tango | *TANG* GO |
| U | Uniform | *YOU* NEE FORM |
| V | Victor | *VIK* TAH |
| W | Whiskey | *WISS* KEY |
| X | X-ray | *ECKS* RAY |
| Y | Yankee | *YANG* KEY |
| Z | Zulu | *ZOO* LOO |

# Time zones

**Table B-2. Time zones.**

Eastern Daylight Time (EDT)
Atlantic Standard Time (AST)

Central Daylight Time (CDT)
Eastern Standard Time (EST)

Mountain Daylight Time (MDT)
Central Standard Time (CST)

Pacific Daylight Time (PDT)
Mountain Standard Time (MST)

Pacific Standard Time (PST)

| UTC | EDT/AST | CDT/EST | MDT/CST | PDT/MST | PST |
|------|---------|---------|---------|---------|------|
| 0000 | 2000 | 1900 | 1800 | 1700 | 1600 |
| 0100 | 2100 | 2000 | 1900 | 1800 | 1700 |
| 0200 | 2200 | 2100 | 2000 | 1900 | 1800 |
| 0300 | 2300 | 2200 | 2100 | 2000 | 1900 |
| 0400 | 0000 | 2300 | 2200 | 2100 | 2000 |
| 0500 | 0100 | 0000 | 2300 | 2200 | 2100 |
| 0600 | 0200 | 0100 | 0000 | 2300 | 2200 |
| 0700 | 0300 | 0200 | 0100 | 0000 | 2300 |
| 0800 | 0400 | 0300 | 0200 | 0100 | 0000 |
| 0900 | 0500 | 0400 | 0300 | 0200 | 0100 |
| 1000 | 0600 | 0500 | 0400 | 0300 | 0200 |
| 1100 | 0700 | 0600 | 0500 | 0400 | 0300 |
| 1200 | 0800 | 0700 | 0600 | 0500 | 0400 |
| 1300 | 0900 | 0800 | 0700 | 0600 | 0500 |
| 1400 | 1000 | 0900 | 0800 | 0700 | 0600 |
| 1500 | 1100 | 1000 | 0900 | 0800 | 0700 |
| 1600 | 1200 | 1100 | 1000 | 0900 | 0800 |
| 1700 | 1300 | 1200 | 1100 | 1000 | 0900 |
| 1800 | 1400 | 1300 | 1200 | 1100 | 1000 |
| 1900 | 1500 | 1400 | 1300 | 1200 | 1100 |
| 2000 | 1600 | 1500 | 1400 | 1300 | 1200 |
| 2100 | 1700 | 1600 | 1500 | 1400 | 1300 |
| 2200 | 1800 | 1700 | 1600 | 1500 | 1400 |
| 2300 | 1900 | 1800 | 1700 | 1600 | 1500 |
| 2400 | 2000 | 1900 | 1800 | 1700 | 1600 |

# Index

## A

absorption wavemeter, 34
ac, 78-94 (*see also* capacitors; inductors)
  circuit components, 81-85
  generator, 179-180
  phase relationship with circuits, 85-87
  sine wave, 180
  study question answers, 298-299
  study questions, 93-94
  waveforms, 78-81 (*see also* sine wave; square wave)
accelerating anode, 196
aeronautical mobile service, 9
aeronautical station, 9
air traffic control (ATC), 206, 210
aircraft stations, 9
  station identification, 25
alternating current (*see* ac)
AM receivers, 128-132
  image reception, 132
  study question answers, 301
  study questions, 133, 135
AM transmitters, 137
  bandwidth of emission, 147-149
ammeter, 34
ampere, 68
ampere-hour (Ah), 75
  calculations, 76

amplifiers, 138-139
  audio, 137
  audio frequency, 129-130, 134
  buffer, 35, 137
  Class A, 138
  Class B, 138
  Class C, 138
  grounded grid, 39
  intermediate frequency, 129, 134
  operational, 42
  power, 137
  radio frequency, 134
amplitude
  average, 80-81
  peak, 79
  peak-to-peak, 79
  root-mean-square (rms), 79-80
amplitude modulation (*see* AM)
AND gate, 199-200
angle modulation, 34
annual inspection, 1-2
anode, 196
  omnidirectional, 42
  stacked, 47
  unidirectional, 48
  vertical, 49
antennas, 128-129, 134, 152-156, 214
  current/voltage relationships, 34
  dummy, 37
  gain, 167

  ground-plane, 154
  Hertz, 153-154
  increasing gain, 155-156
  length, 34
  maintenance, 214
  Marconi, 41, 154-155
  omnidirectional, 42
  resonant frequency, 152-153
  shipboard, 24
  stacked, 47, 156
  study question answers, 304-305
  study questions, 156-157
  unidirectional, 48
  vertical, 49
apparent power, 190
applications, 2
aquadag coating, 196, 215
armature, 179
artificial transmission line, 215
associated ship unit, 2
ATC transponders, 210
Atlantic Standard Time (ATS), 314
atmospheric interference, 35
atom, 65
  copper, 66
  hydrogen, 66
ATR box, 215
attenuator, 215
audio amplifier, 137
audio frequency amplifier, 129-130, 134

# Other bestsellers of related interest

**THE COMPLETE BOOK OF
OSCILLOSCOPES—2nd Edition
—Stan Prentiss**

Save hours of troubleshooting time by mastering top-of-the-line scopes with the professional advice in this book. It reviews the oscilloscope equipment and advanced testing procedures developed during the last five years. You'll appreciate this book's easy-to-read style, logical format, and original photographs taken by video communications expert Stan Prentiss in his electronics laboratory. 320 pages, 165 illustrations. **Book No. 3825, $16.95 paperback, $26.95 hardcover**

**PRACTICAL ANTENNA HANDBOOK
—Joseph J. Carr**

This is the most comprehensive guide available on designing, installing, testing, and using communications antennas. Carr provides a unique combination of theoretical engineering concepts and the kind of practical antenna know-how that comes only from hands-on experience in building and using antennas. He offers extensive information on a variety of antenna types (with construction plans for 16 different types), including high-frequency dipole antennas, microwave antennas, directional beam antennas, and more. 416 pages, 351 illustrations. **Book No. 3270, $21.95 paperback, $32.95 hardcover**

**GREAT SOUND STEREO SPEAKER
MANUAL—with Projects
—David B. Weems**

Beautiful music doesn't have to be painfully expensive. Using this book, you can build your own speaker system, save a lot of money, and get great performance from your stereo equipment. Weems strips the mystery from drivers, crossovers, construction materials, and more to help you design and build the best system for your personal needs. Twelve step-by-step projects are featured, using components from many major speaker manufacturers. 256 pages, 165 illustrations. **Book No. 3274, $16.95 paperback, $25.95 hardcover**

**SOUND SYNTHESIS:
Analog and Digital Techniques
—Terence Thomas**

This authoritative guide gives you access to the most up-to-date methods of sound synthesis—the information and guidance you need to plan, build, test, and debug your own state-of-the-art electronic synthesizer. You'll also find out how to modify or interface an existing unit to gain better synthesized sound reproduction. Packed with diagrams, illustrations, and printed circuit board patterns, this practical construction manual is designed for anyone serious about producing the most advanced synthesized sound possible. 176 pages, 149 illustrations. **Book No. 3276, $14.95 paperback only**

## THE RADIO AMATEUR'S DIGITAL COMMUNICATIONS HANDBOOK
### —Jonathan L. Mayo, KR3T

With the advent of microcomputers and digital communications technology, as well as the new "nocode" operators license, amateur radio is taking off to new levels. This guide is your one-stop reference for all of the major digital modes and their many uses. Here is a complete overview of digital communications for hams, with a discussion of its history, capabilities, and applications, as well as the latest equipment and how to set up a state-of-the-art operational digital communications station. 224 pages, 80 illustrations. **Book No. 3362, $14.95 paperback, $22.95 hardcover**

## PIRATE RADIO STATIONS: Tuning In to Underground Broadcasts
### —Andrew R. Yoder

Tune in, identify, and contact the world of underground radio. Including a complete up-to-date listing of stations, this comprehensive handbook fills the void that has existed in underground radio information for shortwave and longwave hobbyists, DXers, and radio operators. You'll find out how to zero in on even the most unpredictable low-powered stations. 192 pages, 89 illustrations. **Book No. 3268, $12.95 paperback, $19.95 hardcover**

## THE NEW STEREO SOUNDBOOK
### —F. Alton Everest and Ron Streicher

Discover how to optimize your stereo equipment and listening environment. In this book two well respected electronics and audio recording experts bring you an in-depth look at the equipment and techniques that influence sound reproduction, perception, and recording. Easy-to-follow projects, diagrams, tables, and photographs will help you to improve, enhance, and modify your stereo sound. 296 pages, 115 illustrations. **Book No. 3789, $18.95 paperback, $29.95 hardcover**

## BASIC ELECTRONICS THEORY
### —3rd Edition—Delton T. Horn

"All the information needed for a basic understanding of almost any electronic device or circuit . . ." was how *Radio-Electronics* magazine described the previous edition of this now-classic sourcebook. This completely updated and expanded edition provides a resource tool that belongs in a prominent place on every electronics bookshelf. Packed with illustrations, schematics, projects, and experiments, it's a book you won't want to miss! 544 pages, 650 illustrations. **Book No. 3195, $22.95 paperback only**

## THE ILLUSTRATED DICTIONARY OF ELECTRONICS—5th Edition
### —Rufus P. Turner and Stan Gibilisco

This completely revised and updated edition defines more than 27,000 practical electronics terms, acronyms, and abbreviations. Find up-to-date information on basic electronics, computers, mathematics, electricity, communications, and state-of-the-art applications—all discussed in a nontechnical style. The author also includes 360 new definitions and 125 illustrations and diagrams. 736 pages, 650 illustrations. **Book No. 3345, $26.95 paperback, $39.95 hardcover**

## THE MASTER IC COOKBOOK
### —2nd Edition
### —Clayton L. Hallmark and Delton T. Horn

*"A wide range of popular experimenter/hobbyist linear ICs is given in this encyclopedic book."*
### —*Popular Electronics,* on the first edition

Find complete information on memories, audio amplifiers, RF amplifiers, and related devices, in addition to other sections on TTL, CMOS, special-purpose CMOS, and other linear devices. Circuits come complete with pinouts, specifications, and a concise description of the IC and its applications. 576 pages, 390 illustrations. **Book No. 3550, $22.95 paperback, $34.95 hardcover**

## SECRETS OF RF CIRCUIT DESIGN
### —Joseph J. Carr

This book explains in clear, nontechnical language what RF is, how it works, and how it differs from other electromagnetic frequencies. You'll learn the basics of receiver operation, the proper use and repair components in RF circuits, and principles of radio signal propagation from low frequencies to microwave. You'll enjoy experiments that explore such problems as electromagnetic interface. 416 pages, 411 illustrations. **Book No. 3710, $19.95 paperback, $32.95 hardcover**

## THE CET EXAM BOOK—2nd Edition
### —Ron Crow and Dick Glass

An excellent source for update or review, this book includes information on practical mathematics, capacitance and inductance, oscillators and demodulators, meters, dependency logic notation, understanding microprocessors, electronics troubleshooting, and more! Thoroughly practical, it is an essential handbook for preparing for the Associate CET test! 266 pages, 211 illustrations. **Book No. 2950, $13.95 paperback only**

**Prices Subject to Change Without Notice.**

# Look for These and Other TAB Books at Your Local Bookstore

## To Order Call Toll Free 1-800-822-8158
### (in PA, AK, and Canada call 717-794-2191)

or write to TAB Books, Blue Ridge Summit, PA 17294-0840.

| Title | | Product No. | Quantity | Price |
|---|---|---|---|---|
| | | | | |
| | | | | |
| | | | | |
| | | | | |

☐ Check or money order made payable to TAB Books

Charge my ☐ VISA ☐ MasterCard ☐ American Express

Acct. No. _____ Exp. _____

Signature: _____

Name: _____

Address: _____

City: _____

State: _____ Zip: _____

Subtotal $ _____

Postage and Handling
($3.00 in U.S., $5.00 outside U.S.) $ _____

Add applicable state and local
sales tax $ _____

TOTAL $ _____

TAB Books catalog free with purchase; otherwise send $1.00 in check or money order and receive $1.00 credit on your next purchase.

*Orders outside U.S. must pay with international money order in U.S. dollars.*

**TAB Guarantee: If for any reason you are not satisfied with the book(s) you order, simply return it (them) within 15 days and receive a full refund.** **BC**